A Technology Portfolio of Nature Based Solutions

Sean O'Hogain • Liam McCarton

A Technology Portfolio of Nature Based Solutions

Innovations in Water Management

 Springer

Sean O'Hogain
School of Civil & Structural Engineering
Dublin Institute of Technology
Dublin, Ireland

Liam McCarton
School of Civil & Structural Engineering
Dublin Institute of Technology
Dublin, Ireland

ISBN 978-3-030-10347-7   ISBN 978-3-319-73281-7   (eBook)
https://doi.org/10.1007/978-3-319-73281-7

Printed on acid-free paper

This Springer imprint is published by the registered company Springer International Publishing AG part
of Springer Nature.
The registered company address is: Gewerbestrasse 11, 6330 Cham, Switzerland

Foreword

The Nobel Prize-winning atmospheric chemist, Paul Crutzen, has argued we have, since the Industrial Revolution, entered a new geological epoch – the Anthropocene – defined by the overwhelming influence of one species, our own, on the natural world. We are already beginning to witness the catastrophic effects of anthropogenic climate change brought about by the massive disruption of the carbon cycle produced by the emission of greenhouse gases. The significant and unpredictable alterations to the nitrogen, phosphorus and sulphur cycles induced by human action are disrupting both plant and animal life. The terrestrial water cycle, so vital to human existence, has been modified by deforestation, disruptions to river systems and changing land use.

Recent years have seen the food security and livelihoods of millions of men, women and children seriously undermined by unusually severe floods, droughts and storms. Climate change and environmental degradation have enormous and far-reaching consequences on our capacity for the realisation of human rights. Indeed, it is becoming increasingly obvious that the effects of extreme weather events threaten the effective enjoyment of a range of basic human rights, such as the right to safe water and food and the right to health and adequate housing.

Pollution, climate change, the loss of biodiversity and of course the growing scarcity of fresh water are part of a complex system which we must come to understand in each of its parts. It is becoming increasingly apparent that our existing models of production, consumption, distribution and exchange are no longer sufficient to ensure that the needs – in terms of water, food, shelter and intellectual and material resources – of a growing population are met. The great challenge of this century will be to meet the requirement for a just and sustainable development that encompasses and includes all communities and nations without imperilling the ecosystems upon which we all depend.

This will require brave and wise decisions from world leaders and a willingness to re-examine and re-imagine what has become a fractured relationship between the economy, ecology and ethics. It will also require a renewed dedication to the development of science and technology, and a commitment to making new science available to all, to ensure the best outcomes for all of humanity.

At the heart of great and good science lies a creative and moral instinct to explore, to question and to create a better world. Just as the causes of climate change are myriad, so too are its solutions. We must now, from the smallest fragment of the local, think and act globally. These qualities of mind and practice will be essential if we are to, as a national and global community, realise all of the potential of our people and to prove ourselves equal to the difficult challenges that lie ahead in the coming decades.

Will the pursuit of new knowledge and advances in science be adapted to serve the public good, or will it be subordinated to the quest for private profit? This is a pressing question for the development and transfer of technology to and for the developing world and for the achievement of sustainable development and for the mitigation of climate change.

The needs of this new century cannot be met by the stale and disproven assumptions of any inevitability claimed for unilinear change. They will demand new science and new technology, and new modes of thought, animated by and drawing on the spirit of discovery, patience and perseverance.

In its recognition that water is vital human right, and by offering a new model of understanding the consumption and distribution of water, this collection of innovative nature-based solutions represents an important resource for all those seeking to forge a new path to a sustainable future, one that can meet human needs while respecting and potentially restoring ecosystems.

President of Ireland   Michael D. Higgins

Preface

There is a finite amount of water available to the world. The global water cycle involves the circulation of water into and out of various reservoirs, including atmosphere, land, surface water and groundwater. The volume of renewable supplies, through the global water cycle, remains constant, and is set to remain so, despite climate change. This water is not distributed equally as it falls on the earth. Theoretically there is enough water for everyone on the earth, but due to inequality of distribution, certain areas of the world are more water stressed than others. This has resulted in certain parts of the world being well populated, while other parts are scarcely populated.

The universal problem with water has been competition for its use. Circulation of water through the planet has been altered with the reclamation of wetland areas and mangrove forests. The problems with loss of biodiversity, climate change and eutrophication of freshwater and some seas are all consequences of the mismanagement of the water environment.

The traditional approach to water and water infrastructure has been a unidirectional linear model. Water is abstracted at source, treated to potable (drinkable) standard, then used and treated again prior to final disposal to the environment. Water is viewed as a raw material that requires a treatment process to make it a finished product. This is then supplied to users, both domestic and industrial, who use it for various purposes. Some of these uses require advanced treatment to produce water of a higher quality. The use of water reduces the quality and therefore it has to be treated again before it can be discharged back into the environment. Similarly, wastewater treatment systems involve collection, treatment and discharge. This second form of treatment has two purposes; the removal of constituents in the water which could harm the aquatic environment, and the improvement of water quality to raw water standards.

The traditional approach also allows for the transport of water resources over large distances, i.e. from points of storage and capture to centres of population. Simultaneously, rainwater is discharged unused via expensive storm-water drainage systems. Surface water is also seen as a design problem, especially in urban and peri-urban areas. The surface water infrastructure is designed to remove surface water from centres of population, and it can be said that water is designed out of these urban developments.

A **circular economy** is an alternative to a traditional **linear economy.** In a circular economy, resources are kept in use for as long as possible. The maximum value is extracted from them whilst in use, and products and materials are recovered and/or regenerated at the end of each service life. This concept can be applied to water management and is termed "the circular economy of water". This has introduced the concepts of resource recovery and resilience. It is a recognition that the stewardship of water resources up to now has been deeply flawed. In the linear approach to water, it is disposed of after use. The emphasis is on the removal of substances in the water, not on recovery of resources in the water.

The circular economy of water is waste-free and resilient. The circular economy of water sees wastewater and its contents as a potential resource, containing minerals, carbon and nutrients, which can be recovered and reused. This approach requires the development of a smart water economy. By smart, we mean the efficient, innovative and clever use of water. It also involves an increase in the rational use and reuse of water. The past 40 years have seen a growing concern over the status of our water resources. It has seen the almost universal adoption of "The Polluter Pays" principle. It has also seen the application of water pricing as a means of water conservation. Water pricing sees water as a commodity, similar to other commodities, like copper or gold. An economic value is given to the commodity and it is used and traded for profit.

However, water is more than a commodity. It is a basic human right. It is, in fact, a life-support medium. We cannot live without it. High quality water is necessary for health. Industry also cannot function without water, and therefore, economically it is necessary for society to exist. The industrialised society cannot exist without the water that is in commodities or the water that allows the industries to function. Water is also required for energy, and it is an important amenity for recreation and indeed rural income. These functions of life support, industrial use, energy supply and amenity can be termed the *"value of water"*.

The circular economy also asks the question "what is the value in water?" The **"value in water"** can be thought of as the economic and societal value that can be realised by extracting and valorising the resources embedded in used water streams. These include the nutrients, minerals, chemicals, metals and energy which are embedded in water. It also refers to the reuse potential of the water.

Combining these two concepts of a life-support medium, the **value *of* water** and the potential for resource recovery and reuse, or the **value *in* water**, we can talk of the **"worth of water"**. The worth of water is the value of water combined with the value in water. In a circular economy of water how is the worth of water achieved? It occurs when every government decision is governed by water.

This includes integrating water into development planning, building supportive institutional structures with the mandate to control and regulate adaptability in both supply and demand management. It is essential for cities to manage their water as a finite resource that is integral to their overall development planning. It also includes community involvement and community input. However, there are few places in the world where this happens. Water is not given its full worth. To give water its full worth involves change.

The adoption of the principles of the circular economy means we have to *act differently, think differently and interact differently*. Thinking differently means placing the value of water to the forefront of all planning, development and education projects. Water and water stewardship should govern every government decision. It also involves not only recognition of the multiplicity of water sources, e.g. rainwater and saline brackish water, but also the importance of closing water loops by implementing reuse/recycling/cascading and resource and energy recovery. This, as we have seen, involves the value in water.

Interacting differently not only involves governance but most importantly involves participation. It involves a multidisciplinary approach, not only featuring the water-energy-food-land use-climate change network, but also involving cross-disciplinary design and participation, involving architects, engineers, planners, ecologists, government and community stakeholders. It is a bottom-up methodology, rather than a top-down one. These interactions will give rise to new ways of dealing with all users of water be they domestic, agriculture, large urban areas or water transport interests.

Acting differently involves the redesign of the water infrastructure, taking advantage of the recent developments in technology and integrating human-built water infrastructure with nature-based ecosystems.

Nature-based solutions (NBS) are both natural and constructed systems which utilise and reinforce physical, chemical and microbiological treatment processes. These processes form the scientific and engineering principles for water/wastewater treatment and hydraulic infrastructure. Nature-based solutions may be low cost, require low energy for operation and maintenance, generate low environmental impacts and provide added value through the benefits that accrue to humanity (ecosystem services). These benefits include biodiversity, mitigation of the effects of climate change, ecosystem restoration, amenity value and resilience. This new design will result in both centralized and decentralized water technologies which will reduce water loss, promote reuse of water and will feature the use of multiple waters for multiple uses, i.e. "Design Water In". It will also recover resources from these multiple waters. This combination of grey and green water infrastructure will also stimulate sustainability and resilience, which is of great importance in combating climate change events such as droughts and flooding. In the NBS approach nature is integral to the solution, i.e. "*Design Nature In*". In the conventional grey infrastructure approach nature is not integral to the solution, rather the solution is superimposed on the natural surroundings, i.e. "*Design in Nature*". A feature of the circular economy of water and NBS is to "*Design Water In*", whereas the linear economy of water is characterised by removing water for off-site treatment, i.e. "*Design Water Out*".

NBS will form an integral component of future water infrastructure that comprises a mix of high tech human built engineered (Grey) infrastructure and NBS (Green) infrastructure. This combination of approaches can be termed "Hybrid Infrastructure". To achieve this transformation to a smart water society and the integration of the hybrid infrastructure into water design, these innovative technologies require trials and demonstration sites, in effect **living labs**. Living labs are user-centred, innovative ecosystems which aim to integrate research and innovation processes in real-life communities and settings. They place the citizen at the centre of innovation, are constructed in a real-life setting, involve multi-stakeholder participation, a multi-method approach and co-creation, which involves the iterations of designs with different sets of stakeholders. Nature-based solutions are examples of living labs.

This publication aims to define and characterise nature-based solutions (NBS) in terms of water source, contaminants, removal mechanisms and resource recovery potential within the context of a circular economy and to illustrate this definition with a range of case studies. These are selected from members of the European Innovation Partnership (EIP) Water Action Group (AG 228) *"Nature-based technologies for Innovation in water management-NatureWat"*. This action group serves to promote the use of NBS through its technology portfolio, which is based on a number of demonstration sites in the fields of climate change adaption, water and wastewater treatment, resource recovery and reuse, and restoring ecosystems. It will present a multidisciplinary approach to NBS. This approach involves bringing together social scientists, governance representatives, scientists and engineers together with end users to define the problem and opportunities to utilise ecosystem services, i.e. the benefits that accrue to humans from using an NBS methodology.

In the ancient world, technology and nature were combined to solve water problems. Examples of this are the viaducts of the Romans, the Qanats of the Persians and the irrigation systems of the Incas. These NBS allowed the societies to prosper. It would be our hope that nature-based solutions will prove as important a technological contribution to the present issues of climate change, increased population and resource depletion.

Dublin Institute of Technology, Dublin, Ireland   Sean O'Hogain
                                                                   Liam McCarton

Acknowledgements

Thanks are due to a wide range of people: Noreen Layden of the Environmental Sustainability Health Institute and Mark Sweeney of Enterprise Ireland; Prof. Joan Garcia of Universitat Politècnica de Catalunya; Anna Fabregas and Merce Aceves at Area Metropolitiana de Barcelona (AMB); Marc Freixa of Depuradores Osona, Catalunya for site visits; Anna Garfi of Gemma, UPC Barcelona who arranged all for us in Spain – it was not her fault we got lost; Alenka Mubi Zalaznik of LIMNOS in Slovenia, who guided us round her country and the NatureWat sites – thanks for your time and your company; Victor Beumer, Remco Van Ek and Stefan Jansen of Deltares for their time and help; Albert Jansen, of Water Innovation Consulting for inviting DIT to participate in the European Water Supply and Sanitation Technology Platform (WssTP) working groups; John Turner, Aidan Dorgan, Stephen McCabe, Catherine Carson and Richard Tobin in DIT Bolton Street; Catherine McGarvey and Debbie McCarthy in DIT Rathmines: special thanks to Anna Reid, without Anna this would not have been possible – all journeys start from home – to our long-suffering families, thanks, in particular to Jackie, Angie and Sadhbh.

The authors acknowledge the financial support of the Environmental Protection Agency, Ireland.

Contents

Acronyms and Annotations

AdP	*Águas de Portugal*
AMB	Area Metropolitiana de Barcelona
CBA	Cost Benefit Analysis
CBD	Rio Convention on Biological Diversity
CE	Circular Economy
CER	Commission for Energy Regulation
CIREF	Iberian Centre of River Restoration
COD	Chemical Oxygen Demand
CPA	Community Participatory Approach
CW	Constructed Wetlands
DB	Design Build
DBO	Design Build Operate
DBOF	Design Build Operate Finance
DECLG	Department of the Environment, Community and Local Government
DIT	Dublin Institute of Technology
DTC	Development Technology in the Community Research Group
EC	European Commission
EC	European Communities
EcoSan	Eco Sanitation Club
EFRO	European Fund for Regional Development
EIA	Environmental Impact Assessment
EIP	European Innovation Platform
ENSAT	Enhancement of Soil Aquifer Treatment
EPA	Environmental Protection Agency
EQS	Environmental Quality Standards
ERI	Environmental Research Institute
ESHI	Environmental Sustainability Health Institute
GSI	Geological Survey of Ireland
GWP	Global Water Partnership
HLR	Hydraulic Loading Rate
HRT	Hydraulic Retention Time

HWTS	Hybrid Reed and Willow Bed Treatment System
INBO	International Network of Basin Organizations
IRBD	International River Basin Districts
JRC	European Commission's Joint Research Centre
LCA	Lifecycle Analysis
MAR	Managed Aquifer Recharge
NBS	Nature-Based Solutions
NewERA	New Economy and Recovery Authority
NFGWS	National Federation of Group Water Schemes
NIMBY	Not In My Back Yard
NPWS	National Parks and Wildlife Service
NRWMC	National Rural Water Monitoring Committee
OIEAU	Office International de l'eau
OPW	Office of Public Works
P	Phosphate
PE	Population Equivalent
PPCP	Pharmaceuticals Personal Care Products
PTEA	Spanish Water Technology Platform
RWH	Rainwater Harvesting
SDCC	South Dublin County Council
SDRB	Sludge Drying Reed Bed
SF CW	Subsurface
SME	Small Medium Enterprise
STW	Sludge Treatment Wetlands
TSS	Total Suspended Solids
UPC	Universitat Politècnica de Catalunya
WIRC	Water Innovation Research Centre
WssTP	Water Supply and Sanitation Platform
WTP	Water Treatment Plants
WWTP	Wastewater Treatment Plants

List of Figures

List of Tables

Introduction

Organisation of the Book

Chapter 1 proposes a definition of NBS and reviews NBS in terms of the circular economy and proposes a methodology for implementing NBS projects. Chapter 2 discusses the European Innovation Partnership (EIP) on Water, together with the EIP Action Group NatureWat (AG228). The members of the action group are presented together with particular issues which the action group will seek to address. Chapter 3 presents a technology portfolio of nature based solutions. These case studies are based on projects carried out by the members of NatureWat. The case studies are summarised in terms of water sources, contaminants, removal mechanisms and resource recovery potential as well as considering reuse applications. Chapter 4 presents definitions of reclaimed water, reused water and the term "Fit For Purpose". This chapter also presents an overview of potential uses for reclaimed water together with examples of European and global water reuse guidelines. Bottlenecks and barriers related to NBS for water resources management are identified in Chapter 5. Chapter 6 presents case studies at the local, regional and global level. These examples illustrate the application of hybrid infrastructure systems. This is a combination of traditional engineered infrastructure with nature-based solutions.

Chapter 1
Nature-Based Solutions

Abstract This chapter proposes a definition of Nature Based Solutions (NBS), reviews NBS in terms of the circular economy and proposes a methodology for implementing NBS projects. The circular economy of water (CEW) prioritises the concepts of resource recovery and resilience within water resource management. The CEW operating within planetary boundaries, is waste free and resilient and is by design restorative of ecosystems. NBS can form an integral component of this new approach. This publication defines NBS as both natural and constructed systems which utilise and reinforce, physical, chemical and microbiological treatment processes. These processes form the scientific and engineering principles for water/wastewater treatment and hydraulic infrastructure. NBS may be low cost, minimise energy for operation and maintenance, generate low environmental impacts and provide added value through the benefits that accrue to humanity (ecosystem services). These benefits include biodiversity, mitigation of the effects of climate change, ecosystem restoration, amenity value and resilience. This chapter defines and characterises nature based solutions in terms of water source, contaminants, removal mechanisms and resource recovery potential. It will also propose an NBS Methodology.

Keywords Nature Based Solutions · Circular Economy of Water · Ecosystem Services

1.1 Introduction

The methodology of how human society has interacted with the environment has evolved over the last 40 years. Following on from Rachel Carson's work in the 1960s society was concerned with minimising environmental damage (Carson 1962). This had as an underlying principle, the prevention or mitigation of damage to the environment, stated as "do the least possible harm". This gradually led to the adaption, in the 1990s, of the principle of *"sustainability"* and the need to preserve resources and to hand them on intact to future generations (Bruntland Commission 1987). This approach was enshrined within subsequent Environmental Impact Assessment (EIA) procedures (Directive 2014/52/EU). This is the process by which the anticipated effects on the environment of a proposed development or project are measured. If the likely effects are unacceptable, design measures or other relevant mitigation measures can be taken to reduce or avoid those effects.

Fig. 1.1 Characteristics of
the circular economy of
water (CEW)

The current system of water supply and management is based on a linear approach, focusing on commodity sourcing, treating, using and disposing. Currently, water demand is typically met by importing large volumes of water across long distances from neighbouring catchments. Simultaneously, rainwater is discharged unused via expensive storm water drainage systems. Similarly, wastewater treatment systems involve collection, treatment and discharge. This contrasts with the objectives of the circular economy as described on the next section.

1.2   The Circular Economy of Water (CEW)

The circular economy has introduced, in the last few years, the concepts of resource recovery and resilience. The circular economy is by design restorative of ecosystems. In the linear approach to water, products are disposed of after use. The circular economy, operating within planetary boundaries, is waste free and resilient. The circular economy of water (CEW) sees water and its contents, as a resource (Fig. 1.1).

1.3   Nature-Based Solutions

The term 'nature-based solutions' (NBS) has been adopted to inform policy and discussion on biodiversity and conservation, climate change adaptation, and the sustainable use of natural resources (Potschin et al. 2015). The term NBS appears to have first been used in the early 2000s, in the context of solutions to agricultural problems. NBS has also been used in discussions on land-use management and planning and water resource management, i.e. the use of wetlands for wastewater treatment and the value of harnessing ecosystem services from wetlands as a form of nature-based solution for watershed management (Guo et al. 2000; Kayser and Kunst 2002; Brink et al. 2012). The NBS concept was also used to describe industrial design and biomimicry. The term "biomimicry" has also been used for green infrastructure and other soft engineering approaches, which have been used as nature-based solutions to urban water management problems. Here the term refers to learning from nature, rather than finding strategies based on nature that would contribute to its conservation (Grant 2012).

More recently NBS have been selected as a priority area for the European Commission (EC) Horizon 2020 Research Programme, though more than one definition of NBS can be found in related literature. The EC Expert Group on NBS suggests that the NBS concept *"builds on and supports other closely related concepts, such as the ecosystem approach, ecosystem services, ecosystem-based adaptation/mitigation, and green and blue infrastructure"* (EC 2015). Another report for Horizon 2020's Societal Challenge 5 (EC 2014) proposes that NBS and the utilisation of biomimicry be used to position the EU as a world leader in the development of industrial and technological solutions *"inspired by, using, copying from or assisted by nature"*. This idea is also included in the aforementioned EC Expert Group Report on NBS definition as follows: *"NBS therefore involve the innovative application of knowledge about nature, inspired and supported by nature"* (EC 2015). It is further stated in the report that industrial challenges and environmental problems caused by human activities can be resolved *"by looking to nature for design and process knowledge"*, but these aspects are not strongly emphasised. The EU BiodivERsA (www.biodiversa.org) also view NBS as being a way to *"conserve and use biodiversity in a sustainable manner"* (Balian 2014). There are, however, some differences in emphasis on the components and aims of NBS.

These different perspectives are largely compatible. However, what is not clear, is how NBS differs from other concepts associated with improving human wellbeing, i.e. by managing ecosystem services and natural capital in appropriate ways. Yet, a clear link between NBS and these concepts is needed to ensure consistency and avoid redundancy or confusion.

When NBS are considered from a water management viewpoint, and with the focus on natural technologies and systems that replicate scientific and engineering principles, the following definition can be proposed. This links ecosystem services, natural capital and NBS.

The authors propose the following definition:

> *Nature-based solutions are both natural and constructed systems, which utilise and reinforce, physical, chemical and microbiological treatment processes. These processes form the scientific and engineering principles for water/wastewater treatment and hydraulic infrastructure. Nature based solutions may be low cost, require low energy for operation and maintenance, generate low environmental impacts and provide added value through the benefits that accrue to humanity (ecosystem services). These benefits include biodiversity, mitigation of the effects of climate change, ecosystem restoration, amenity value and resilience.*

Fig. 1.2 New approach
required to implement a
NBS

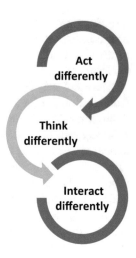

1.4 Are There Nature-Based Solutions?

Nature – can be considered as relating to biodiversity as a totality or the individual elements of biodiversity (individual species, habitats, ecosystems), and/or ecosystem services.

Nature-based– can be considered as referring to ecosystem approaches, ecosystem-based approaches, biomimicry, or direct utilisation of elements of biodiversity.

Solutions –recognisable solutions to a specific problem or challenge.

It is the latter term that distinguishes the NBS approach from other previous terminology, such as sustainable solutions and resilience. When responding to a challenge in the past, the normal approach was to define the problem being addressed. This involved, understanding the context, and then reviewing the technological solutions available. This approach often led to a single focused technological solution. The proposal of a nature-based solution requires that the problem be solved using a multidisciplinary approach (Potschin et al. 2015). The innovation supplied by the nature-based approach is that the question that is addressed may not have a purely technological solution. The review of possible alternative solutions start with the question '**is there a nature-based solution?**'. Thus, the field of possible solutions and the range of options considered are broadened. This then facilitates exploring a NBS centred design methodology.

Problem solvers or opportunity finders
A review of EIP case studies has led to the conclusion that to initiate and promote NBS, a change has to be made in the way we act, the way we think and the way we interact when considering water infrastructure projects (De Vriend and Van Koningsveld 2012). Figure 1.2 illustrates this approach graphically.

Act Differently

To effect a change in how we act, and to facilitate using NBS, it is necessary to consider the context of the project not only in terms of the physical site (both biotic and abiotic), but also in terms of the socio-economic and the governance issues surrounding the problem. This approach, which also takes into account the context as an open ecosystem, is in marked contrast to the traditional problem solving approach followed by project designers, which tends to focus on a single aspect (technological).

The traditional approach can be said to focus on function and to solve a narrowly defined problem in a given timeframe and for a given cost. This traditional method, best described as linear, sought to first define the problem, before progressing to review and propose alternative solutions. These alternatives would then be evaluated using such metrics as EIA, Lifecycle Analysis (LCA), Cost Benefit Analysis (CBA) and others. This method produced a preferred solution. If there was no solution forthcoming, designers returned to defining the problem and proceeded as before until a solution was reached.

In following an NBS methodology, which is a circular approach, the context of the project is dealt with by adopting a multidisciplinary outlook from the beginning. The multidisciplinary approach involves bringing together social scientists, governance representatives, scientists and engineers together with end users to define the problem. This management group then define multi-functional opportunities within the context of the project. These opportunities are also referred to as ecosystem services, as they are the benefits that accrue to humans from using an NBS methodology. These opportunities not only solve the engineering problems but also supply added value. This added value is typically given in terms of ecosystem services. These benefits can include any or all of the following:

- Adaption to climate change,
- Wastewater treatment,
- Ecosystem restoration or resource recovery,
- Biodiversity,
- Recreational amenities.

Think Differently

NBS not only deliver the primary functions for which the project was designed, but also provide added value from both an ecological and economic perspective. For example, the issue of flooding in a particular catchment might be defined by a technical review which defines the problem as one of limited capacity within a river system for certain storm events. The solution may focus on methods of online or offline storage and may proceed to evaluate and rank the possible solutions in terms of Environmental Impact assessment (EIA), Lifecycle Analysis (LCA), Cost Benefit Analysis (CBA) and others. Prerequisites, such as budget and time constraints, often narrow the scope of a project and preclude or hamper innovative solutions. The preferred solution, may be the most technically feasible to solve the narrow problem (increase storage locally) with least environmental impact and minimal cost. Adopting an NBS methodology may widen the scope of the project and offer new perspectives and opportunities. The issue of flood protection may be seen to offer possibilities to create new habitats. The example of the Green Gate project in Rotterdam illustrates the possibility of combining engineering solutions with ecosystems for bank protection and ecosystem services (Deltares 2015a).

A change in thinking involves incorporating the characteristics of NBS from the start. These characteristics include:

1. **Considering multifunctional solutions.** This may involve catering for more than one function in a project and therefore extending traditional proven design approaches using dynamic natural or environmental processes.
2. **Considering the project as a dynamic entity** that is in flux and open to change. Natural processes are not static. Therefore resilience has to be built in. Though the project may be built in a natural setting, i.e. **building in nature**, the change in thinking involves **building nature in**.
3. **Addressing the level of uncertainty** that is increased when dynamics and multi functions are considered. Natural systems involve the introduction of uncertainty and may increase some levels of risk. Uncertainty can be allowed for and dealt with by a knowledge base, which increases the available information. However contingency measures and flexibility are required as built-in adaptive measures to increase the feasibility of the solution.
4. Incorporating the increases in risk that follow on from dynamic and natural systems. Such concepts as uncertainty are what mainstream project designers seek to avoid and the **idea of learning by doing**, which is an underlying principle of NBS, is not widely accepted (Deltares 2015b).

The European Innovation Partnership (EIP) action group NatureWat was set up to promote NBS, and to make available a knowledge base on various NBS technologies. This group has a portfolio of NBS, which aim to make available the NBS technologies and methodologies. This technology portfolio consists of demonstration plants, which while serving to supply ecological and economic services, also function to further the understanding of how to best implement a NBS. *These demonstration plants are tactile, practical and easily accessible in terms of access to the plant and its environs but also in access to the technology used and the scientific and engineering principles underlying the technology. They serve to promote the NBS approach by demonstrating how the problem was identified and how the solution was arrived at. The demonstration sites also illustrate the NBS methodology.*

Interact Differently
To effect a NBS methodology, a change is required in how we interact, and this requires interdisciplinary collaboration and active stakeholder involvement (De Vriend and Van Koningsveld 2012). Water-related infrastructure projects are likely

to affect the interests of a variety of stakeholders, especially in densely populated areas. **"Building Nature In"** also means building with society. Stakeholder involvement is important for two reasons:

- Traditional infrastructure projects often encounter growing resistance from people who will be affected by the project. It is easy to dismiss such resistance as the "Not In My Backyard", or NIMBY syndrome. However, project developers have to recognise that they are interfering with these people's social habitats.
- Local people know a lot about the area where they live, and their knowledge base can be very useful for understanding natural systems and processes, and how they will interact with hard engineering structures. Stakeholder involvement can inspire surprising new solutions. Involving the public provides valuable insights into local systems and processes, and so is more likely to lead to better solutions that stakeholders are more likely to accept. Rather than opposing ideas that have been precooked in some faraway 'ivory tower', people take ownership of projects and even promote them. Therefore the interaction could be summed up as the community participatory approach (CPA), where the community is involved in all aspects of the project.

There is also a need to develop a *"hybrid engineer"*. This is an engineer who has a background in social science, ecology and environmental services. Such individuals, and they can also be hybrid architects and hybrid planners, allow a greater nature-based input as a result of their training and experience in green projects. The inclusion of legislators and governance has been mentioned. Such flexibility can also be incorporated into the design and build stage of the project or into such other existing procurement methods such as Design Build (DB), Design Build Operate (DBO) and Design Build Operate Finance (DBOF). Further innovations may involve management and operation. It is essential that the primary function of infrastructure be aligned with the interests of both nature and stakeholders, in order to arrive at sustainable and socially acceptable solutions.

1.5   Towards a Nature-Based Solutions Methodology

NBS challenges project developers, designers and users to think, act and interact differently. Each project provides a unique opportunity to induce positive change and NBS can be introduced in any phase of any project. The case studies in Chapter 3 describe projects that have been realized using NBS. These projects taken together form a knowledge base of NBS systems.

They also serve to suggest an NBS methodology and taken together with other studies can assist in drawing up a set of principles for NBS project implementation (De Vriend and Van Koningsveld 2012):

1. **Understand the context of the problem/project**. This stage differs from conventional engineering analysis in that it involves a multidisciplinary consultation group made up of engineers and non-engineers including stakeholders. The problem is evaluated in a holistic manner from viewpoints of the many disciplines involved in the project. This includes the environmental, technical, societal and aesthetic aspects of the project. This involves identifying ecosystem services, potential and actual.
2. **Identify realistic alternative solutions** that where possible, use NBS or that provide or use ecosystem services.
3. **Evaluate each alternative**, from an engineering and ecosystem point of view and format a multifaceted solution yielding added value.
4. **Consider** the proposed NBS design analysis in terms of practical limitations and governance. Fine tune where necessary.
5. **Finalise Initial Design Phase** – prepare the solution for implementation in the next phase of the project.

The general design process may be approached from the perspectives below:

The natural environment perspective
In any project, opportunities for NBS are to be found in the natural environment or ecosystem in which the project is to be embedded. Each environment is unique, with its own characteristics, related ecosystem services and associated opportunities.

The project perspective
Each phase of a project presents an opportunity to introduce NBS. Project phases include: initiation, planning and design, construction, and operation and maintenance.

The governance perspective
The governance context, involves the complex set of legislation, regulations, decision-making processes, etc. It also involves networks, regulatory contexts, knowledge contexts and realization frameworks.

The knowledge base
The knowledge base, consists of a wide range of tools, demonstration sites, case studies and other examples. The tools include methods, concepts and strategies that can be used in the different project phases and design steps. Together, the example cases form a technology portfolio of NBS as they have been implemented in projects. The knowledge pages contain information on the various topics and issues that have been addressed during the programme.

1.6   Further Information

EcoShape is a consortium of Dutch companies that include international dredging contractors, public bodies and engineering firms and research institutes such as Deltares. They have developed course materials and tutorials that are being used in workshops and training courses at various collaborating education institutes, i.e. Delft University of Technology, Wageningen University and Research Centre, and the Zeeland and Van Hall Larenstein Universities of Applied Sciences (www.ecoshape.nl).

Chapter 2
European Innovation Platform Action Group NatureWat (AG228)

Abstract This chapter discusses the European Innovation Partnership (EIP), together with the EIP Action Group NatureWat. Almost half of European freshwater bodies are currently not achieving the good ecological status set by the EU Water Framework Directive. Water scarcity, droughts and floods are an increasingly frequent and wide-spread phenomenon in the European and non-European countries. In this scenario, NBS are proven to be cost-effective solutions for wastewater treatment, climate change mitigation, disaster risk reduction, flood protection, greening cities, degraded areas restoration and biodiversity preservation. Their success is related to good performance, potential low maintenance and operational costs, minimised energy requirements, resulting in improved environmental and public health. NatureWat has been formally adopted as an action group within the EIP Water structure. The members of the action group are presented together with particular issues which the action group will seek to address. This Action Group aims to identify and overcome bottlenecks and barriers (e.g. market opportunities, policy implementation, financial issues and technical aspects) related to nature-based solutions for water resources management in rural, peri-urban and urban areas of European and non-European countries. The final goal is to define innovative marketable technologies addressing water challenges related to ecosystem services (www.eip-water.eu/NatureWat).

Keywords European Innovation Partnership · NatureWat Action Group

2.1 European Innovation Partnership (EIP)

www.eip-water.eu

The European Innovation Partnership (EIP) on Water facilitates the development of innovative solutions to address major European and global water challenges. The objective is to pool expertise and resources by bringing together public and private actors at EU, national and regional level, combining supply- and demand-side measures. The EIP Water also supports the creation of market opportunities for these innovations, both inside and outside of Europe. The EIP Water aims to advance the European knowledge base for innovations within the water sector across the public and private sector, non-governmental organisations and the general public. At the core of the EIP Water and its implementation are a number of multi-stake-holders Action Groups (AGs). These groups are composed of a large variety of research institutions, small and medium sized enterprises (SMEs), public institutions and others. Activities within these action groups are being developed to over-come five key barriers to innovation in water management: Finance, Procurement, Partnerships, Regulation and Demonstration Sites.

2.2   NatureWat Action Group (AG 228): *"Nature-Based Technologies for Innovation in Water Management"*

Almost half of European freshwater bodies are currently not accomplishing the good ecological status set by the EU Water Framework Directive (Council Directive 2000/60/EC). Water scarcity, droughts and floods are an increasingly frequent and widespread phenomenon in the European and non-European countries. In this scenario, NBS are proven to be cost-effective solutions for wastewater treatment, climate change mitigation, risk disaster reduction, flood protection, greening cities, degraded areas restoration and biodiversity preservation. Their success is related to good performance, low maintenance and operation costs, low or lack of energy requirements, resulting in improved environmental and public health.

NatureWat has been formally adopted as an action group within the EIP Water structure. This Action Group aims to identify and overcome bottlenecks and barriers (e.g. market opportunities, policy implementation, financial issues and technical aspects) related to nature-based solutions for water resources management in rural, peri-urban and urban areas of European and non-European countries. The final goal is to define innovative marketable technologies addressing water challenges related to ecosystem services (www.eip-water.eu/NatureWat).

2.3   NatureWat Action Group Members

The action group members comprise a cross section of research institutes, government and non-government organisations and professional firms working on developing nature-based solutions for a variety of water management issues across Europe. The Action Group promoter and Coordinator is Professor Joan Garcia, Universitat Politècnica de Catalunya (UPC), Barcelona, Spain. Figure 2.1 illustrates the existing demonstration sites and action group partners.

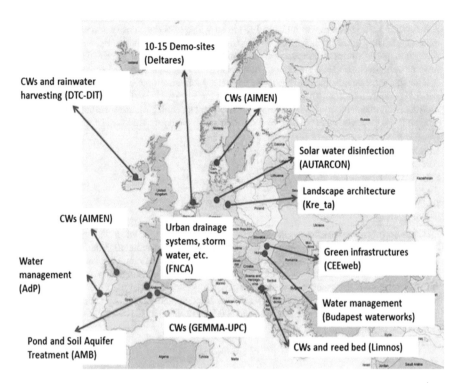

Fig. 2.1 NatureWat members

Universitat Politècnica de Catalunya BarcelonaTech. Environmental Engineering and Microbiology Research Group (GEMMA-UPC), Spain.	 UNIVERSITAT POLITÈCNICA DE CATALUNYA BARCELONATECH Group of Environmental Engineering and Microbiology www.gemma.upc.edu

GEMMA-UPC has long term experience in applied research of NBS for water and wastewater treatment, water reuse and risk prevention. GEMMA has had an active role in boosting and implementing these technologies at international level, including developing countries. The main research topics are:

1. Natural low-cost bioprocesses for wastewater and sludge treatment;
2. Wastewater treatment with algal based culture and biofuel production from algal biomass;
3. Microbial fuel cells in constructed wetlands;
4. Phosphorous removal and recovery from wastewater;
5. Life cycle assessment and economic evaluation of products and technologies;
6. Numerical simulation of bioprocesses.

AIMEN Technology Centre, Spain.	 www.aimen.es

(continued)

AIMEN Technology Centre was set up in 1967 as an initiative of the industry as a non-profit private association. The Centre is focused on developing and strengthening the competitive capacities of companies through the promotion and execution of R&D activities, as well as providing technological services of high added value. The goal of AIMEN is to be a technological and strategic partner, thus contributing to the improvement of their technological capabilities and increasing their competitiveness. AIMEN provides industry with technological services and engages in R&D activities in different areas such as environmental technologies, laser processing, joint technologies, materials and manufacturing processes, engineering, industrial design, simulation and automatics or industrial organization.

Development Technology in the Community Research Group (DTC-DIT), School of Civil and Structural Engineering, Dublin Institute of Technology, Bolton Street, Ireland.	 www.dit.ie/dtc

DTC has been active in research in the areas of water resources, wastewater management, resilient and sustainable water supplies and appropriate technology applications in water self-sufficiency, both in Ireland and internationally. In Ireland, this has focused on rainwater harvesting systems which were constructed, installed, and monitored by DTC. Waste management research has been with Hybrid Reed beds, willow bed wastewater polishing systems, investigations into zero discharge waste treatment systems and also sludge treatment. Internationally DTC have been involved with water and wastewater projects in: Sierra Leone 2009–2012, Bolivia 2011 and Liberia 2014.

Deltares Enabling Delta Life, The Netherlands	 www.deltares.nl

Deltares is an independent institute for applied research in the field of water, subsurface and infrastructure. Throughout the world, they work on problem analysis and smart solutions, innovations and applications for people, environment and society. Their main focus is on deltas, coastal regions, cities and river basins.

Their Nature-Based Engineering programme offers new solutions for flood protection or prevention, wastewater treatment and water storage in intensively-used deltas. Involving nature in the process makes it possible, for example, to improve flood defenses and generate societal benefits. The flood defenses are under considerable pressure due to phenomena such as sea-level rise, land subsidence and periods of extreme rainfall. Deltares develops top-end knowledge and tools in the fields of planning, designing, installing and operating of Nature-based Solutions.

Barcelona Metropolitan Area (AMB), Spain	 http://www.amb.cat

AMB is a public administration composed of 36 municipalities and 3.2 million inhabitants. The main competencies include social cohesion, territorial and urban planning, mobility, transport, waste management, water cycle management, environment protection, social housing. With regard to water cycle, the main responsibilities include drinking water supply, wastewater treatment and reuse. AMB has been involved in a pilot project about soil aquifer treatment to recharge Llobregat River and in a project for flood risk prevention by ponds implementation.

AdP – Águas de Portugal, SGPS, Portugal	 http://www.adp.pt

AdP Group develops R&D activities, in partnership with other institutions, companies and Universities, in a wide range of subjects including novel processes and methods for optimizing water network systems and wastewater treatment processes and developing simulation tools for optimizing wastewater treatment and collection. AdP and its subsidiary companies work together, in matters of water supply and wastewater treatment. They cover about 80% of Portugal population and operate/manage (2011 data) 899 Wastewater Treatment Plants (WWTP), 2187 pumping stations, 6347 km of main sewage system, 845 water abstraction, 247 Water Treatment Plants (WTP), 1361 water reservoirs and 12,520 water main supply network and 485 Mm^3/year of wastewater treated and 611 Mm^3/year of produced drinking water.

Budapest Waterworks, Hungary

BUDAPEST WATERWORKS

http://vizmuvek.hu/en

Budapest Waterworks, being the utility market leader of the Hungarian water sector, has a history of 147 years providing high quality drinking water to more than two million inhabitants of the capital city and the surrounding settlements. Budapest Waterworks is a constantly improving operator with strong interests in the field of ecological water and wastewater management, ecosystem services and zero emission processes. Through strategic partnerships and cluster memberships the company plays a significant role in the international water and wastewater sector.

LIMNOS Ltd., Slovenia

http://www.limnos.si

LIMNOS carries out the following activities/services: development and design of wastewater and sludge treatment, natural reclamation of degraded areas (including landfill sites), revitalization of streams and lake management (including eutrophication prevention). Their experiences include over 200 built constructed wetlands for treatment of domestic sewage, industrial wastewaters and landfill leachate and several projects of sludge sanitation from WWTP using sludge drying reed bed technology (SDRB). LIMNOS is also involved in the development of a method for degraded soils rehabilitation, which was tested in one major landfill site in Slovenia. It is involved in measures for reducing and rehabilitation of risk factors for eutrophication in five locations in Slovenia. Current R&D efforts include optimisation of drying reed bed technology.

AUTARCON GmbH, Germany

www.autarcon.com

AUTARCON is specialized in decentralized water treatment technologies that can be run without the external supply of energy and chemicals. The core component of the system is an inline electrolytic unit to disinfect the water for a pathogen free and safe supply. These disinfection units are installed worldwide, with a current focus on India and Africa. AUTARCON is conducting on-going research to develop products, which allow a stronger market penetration by offering modularized water treatment solutions that can be installed depending on locally existing water quality challenges. This way AUTARCON can offer cost and energy efficient solutions for polishing treated wastewater, as well as the removal of turbidity, pathogens, hardness, iron and manganese from source water.

(continued)

Kre_ta Landschaftsarchitektur, Germany	www.kreta-berlin.de

Kre_Ta Landscape Architecture is a creative small medium enterprise (SME), running an office for landscape architecture and urban planning in Berlin, Germany. Established in 2003 Kre_Ta is focusing on sustainable and green urban designs with a strong focus on R&D projects to develop and design eco-innovative solutions. The Kre_Ta staff has a wide range of experiences in ecological architecture, urban design, and R&D activities in the fields of urban and space planning and rain water harvesting concepts in cities.

Fundación Nueva Cultura del Agua (New Water Culture Foundation), Spain	www.fnca.eu/en

FNCA is an Iberian (Spain and Portugal) non-profit organization composed of over 200 outstanding members from academia, research institutions, public administration, private sector, stakeholders and citizens, aiming at promoting a change towards a more sustainable water management. Main skills: wetlands, rivers and riverbank restoration, green river infrastructures for flood risk management; methods for estimation of environmental flows, assessment of river habitat; biological indicators; application of inter and trans-disciplinary approaches; models for integrating the ecological, economic, social and cultural dimensions; water governance and participatory approaches.

CEEweb for Biodiversity Non-Governmental Organisation, Hungary	 www.ceeweb.org

CEEweb for Biodiversity is a network of non-governmental organizations in the Central and Eastern European region. The mission of CEEweb is the conservation of biodiversity through the promotion of sustainable development. CEEweb follows water-related EU policy developments and raises awareness regarding the possibilities to influence it, with a specific website for Green Infrastructure, which also includes Blue Infrastructure. They update different stakeholders about latest issues and present case studies as examples.

2.4 Specific Issues Which the NatureWat Action Group Seeks to Address

The currently identified challenges which the action group members will seek to address are:

2.4.1 Barriers to Market Uptake

(a) To improve technical performance of nature-based technologies for water management (i.e. artificial wetlands, ponds, sustainable urban drainage systems, combined sewer overflow treatment systems, green roofs, vertical gardens) by defining innovative design and operation appropriate in different environmental and socio-economic conditions and for different purposes (i.e. wastewater treatment, climate change mitigation, risk disaster reduction, flood protection, greening cities, degraded areas restoration and biodiversity preservation).

(b) To increase market opportunities for nature-based technologies in water management by identifying innovative and successful strategies for market uptake and verifying the show cases where market strategies are implemented.

(c) To increase the positive perception and social acceptance of nature-based technologies for water management through the dissemination of successful show cases among public entities, companies, users and associations.

(d) To improve networking and collaboration among European and international stakeholders (i.e.: R + D entities, Universities, SMEs, municipalities) interested in this topic.

2.4.2 Demand/Market Potential

As stated in the EC Roadmap on Resources Efficiency (EC 2011) 60% of the Earth's ecosystem services have been degraded in the last 50 years. It was declared that by 2020, natural capital and ecosystem services should be properly valued and accounted for by public authorities and businesses. NBS are identified as instruments for investing in ecosystems. NBS for water management are considered efficient strategies for wastewater treatment, climate change mitigation, risk disaster reduction, flood protection, greening cities, degraded areas restoration and biodiversity preservation. They are proven to be cost-effective solutions which require low energy and low operation and maintenance. Since NBS for water management can be applied to urban, peri-urban and rural areas of developed or emerging countries to address different urgent challenges, their market potential is huge. NatureWat aims to quantify market opportunity and identify strategies for market uptake. Solutions proposed by NatureWat will have an impact on an international market, creating jobs and stimulating the green economy.

2.5 European Working Groups

NatureWat members work across a variety of key networks, platforms and European working groups dedicated to NBS for water management. A brief synopsis is presented in Table 2.1.

Table 2.1 NatureWat European working groups (WGs) and networks

Water supply and sanitation Technology Platform (WssTP – EU Water Platform)	
WssTP is the European Technology Platform for Water. Initiated by the European Commission in 2004, WssTP strives to promote coordination and collaboration of Research and Innovation in the European water sector, improving its competitiveness.	
Website	http://wsstp.eu/
Working Group on Green Infrastructure of the WssTP (EU Water Platform) (lead by Deltares)	
The main objective of this WG is to promote the development of innovative and effective green infrastructure solutions based on scientific evidence for local, regional, river basin and coastal environments.	
Website	http://wsstp.eu/communities/working-groups/
Spanish Water Technology Platform (PTEA)	
The Spanish Water Technological Platform is an association integrated by over 100 companies, universities and research institutions from Spain, with the aims of expanding knowledge on water, provide networking opportunities and relations among different agents in the water sector and research funding opportunities.	
Website	http://www.plataformaagua.org/
Catalan water network (Water.cat)	
The network was created to give more visibility to the partners at international level. The aim was to communicate Research & Development requests and challenges directly to the European Commission, in order to achieve funding instruments in programs financed by the European Union and to be represented in forums and European lobbying and decision making groups.	
Eco Sanitation Club (EcoSan Club)	
The EcoSan Club was funded as a non-profit association in 2002 by a group of people active in research and development as well as planning consultancy in the field of sanitation. The underlying aim is the realisation of ecological concepts to close material cycles in settlements.	
Website	http://www.ecosan.at/
International Network of Basin Organizations (INBO)	
The main goal of the International Network of Basin Organizations is to upgrade and support the development of organizational initiatives for integrated water resources management (IWRM) in river basins/lake basins/aquifer level.	
Website	www.inbo-news.org
Office international de l'eau (OIEAU)	
The objective of the International Office for Water is to gather public and private partners involved in water resources management and protection in France, Europe and the rest of the world. Its mission includes: Training, Information, Management and Cooperation in the water sector.	
Website	www.oieau.fr
Global water partnership (GWP)	
GWP was founded in 1996 to foster integrated water resources management (IWRM) which is defined as the coordinated development and management of water, land, and related resources in order to maximise economic and social welfare without compromising the sustainability of vital environmental systems. Their mission is to advance governance and management of water resources for sustainable and equitable development.	
LIMNOS is a national partner (contact office) for GWP Slovenia. This year they finalised a small water retention measure project: http://www.gwp.org/en/GWP-CEE/gwp-cee-in-action/news-and-activities/Small-water-retention-measures/	

(continued)

Table 2.1 (continued)

Website	http://www.gwp.org/en/

International Rivers

International Rivers works to protect rivers and rights and to promote green and soft solutions for meeting water, energy and flood management needs around the world. They have experience in dam removal, soft solutions for floods risk management and environmental flows.

Website	www.internationalrivers.org

Iberian Centre of River Restoration – CIREF

CIREF is a Spanish entity composed of experts in river restoration coming from universities, public administrations, consulting companies and NGOs. CIREF is promoting NBS for river management, flood prevention and river biodiversity enhancement. CIREF is linked to the **European Centre for River Restoration** (ECRR), following the examples of countries as Denmark (DCVR), Italy (CIRF) or United Kingdom (RRC).

Website	http://www.cirefluvial.com/

CEEweb for Biodiversity

CEEweb deals in its Working Groups with Blue-Green Infrastructure, NBS, MAES, Ecosystem Services and Natura 2000, among others.

Website	www.ceeweb.org

European Habitats Forum WG on Target 2 (Green Infrastructure Implementation and Restoration)

The WG comprises leading European NGOs active on the Topic of Target 2 of the European 2020 biodiversity Strategy in the policy context.

Chapter 3
Nature-Based Solutions: Technology Portfolio

Abstract This chapter collates best National and International Practice through a series of case studies. These case studies are based on demonstration projects carried out by the members of NatureWat. The NatureWat knowledge base consists of a wide range of tools, demonstration sites, case studies and examples. These demonstration plants are tactile and easily accessible in terms of access to the plant and its environs but also in access to the technology used and the scientific and engineering principles underlying the technology. The tools developed include methods, concepts and strategies that can be used in the different project phases and design steps. Together, the example cases form a technology portfolio of NBS. The concept of NatureWat is to increase the knowledge base continuously and enhance the concept of NBS by constantly evaluating the effectiveness across a variety of applications. The case studies are summarised in terms of water sources, contaminants, removal mechanisms and resource recovery potential as well as considering reuse applications.

Keywords Living Labs · Nature Based Solutions Technology Portfolio

© Springer International Publishing AG, part of Springer Nature 2018   21
S. O'Hogain, L. McCarton, *A Technology Portfolio of Nature Based Solutions*,
https://doi.org/10.1007/978-3-319-73281-7_3

3.1   NBS Technology Portfolio

The EIP action group NatureWat was set up to promote NBS. The NatureWat knowledge base consists of a wide range of tools, demonstration sites, case studies and examples. These demonstration plants are tactile and easily accessible in terms of access to the plant and its environs but also in access to the technology used and the scientific and engineering principles underlying the technology. The tools developed include methods, concepts and strategies that can be used in the different project phases and design steps. Together, the example cases form a technology portfolio of NBS. The concept of NatureWat is to increase the knowledge base continuously and enhance the concept of NBS by constantly evaluating the effectiveness across a variety of applications.

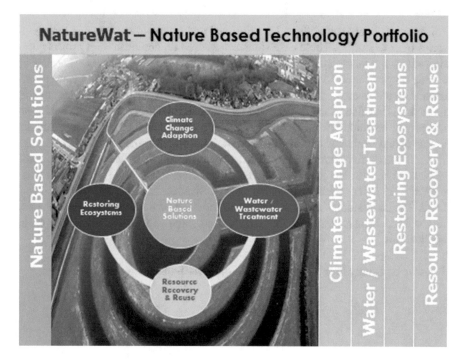

3.2   Case Study 1: Inland Shore Concept Lake Ijsselmeer

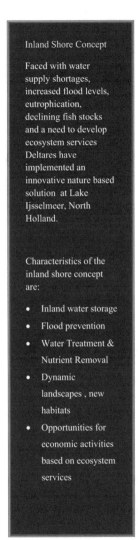

Inland Shore Concept

Faced with water supply shortages, increased flood levels, eutrophication, declining fish stocks and a need to develop ecosystem services Deltares have implemented an innovative nature based solution at Lake Ijsselmeer, North Holland.

Characteristics of the inland shore concept are:

- Inland water storage
- Flood prevention
- Water Treatment & Nutrient Removal
- Dynamic landscapes , new habitats
- Opportunities for economic activities based on ecosystem services

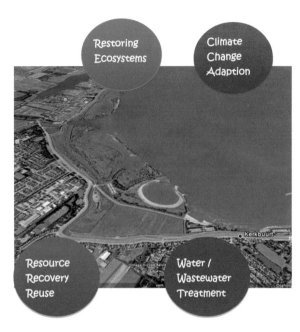

www.deltares.nl **Deltares** Enabling Delta Life **DeltaresThe Netherlands**	**Site Location**: Lake Ijsselmeer, Koopmanspolder, North Holland	**Project Partners** Deltares, Rijkswaterstaat, Staatsbosbeheer, Hoogheemraadschap Holland. Hoogheemraadschap Hollands NoorderkwartierProvince Noord-

Fig. 3.1 Inland shore concept

3.2.1 Site Location:Koopmanspolder

The Inland shore project is located at Koopmanspolder near the village of Andijk along the border of Lake Ijsselmeer, in the Province of North-Holland in the Netherlands. It receives water from Lake Ijssel, has a surface area of 1,100 km^2 which serves as a reservoir and a source of drinking water, as well as an area of recreation. Koopmanspolder lies 1.5 m below lake level and has a surface area of 16,00 m^2 and was formerly a depot for soil storage.

3.2.2 Inland Shore Concept

In 2010, Rijkswaterstaat, the Dutch water body and Deltares joined forces with the Province of Noord-Holland, the regional waterboard, Hoogheemraadschap Hollands Noorderkwartier, and the Government Service for Land and Water Management (DLG) to develop a unique project: Koopmanspolder is the first pilot project within the broader innovative framework of "inland shores". It aims at creating a ring of high quality wetland areas around Lake IJsselmeer where ecosystem services are optimised for different land-use functions. Figure 3.1 illustrates the concept of an Inland shore. This is an area for water storage connected to a nearby lake or river, in which ecosystem services are optimised for multiple land-use, thereby creating new economic opportunities. Large water level fluctuations are taken into account in advance, thus optimising sustainability. The additional area provides room for water storage, facilitating flood prevention and reducing susceptibility to water scarcity. Inland shores are connected to the lake or river through inlets in the dike. The design concept allows the storage area to function as a helophyte (plant which has its buds underwater) filter to allow improvement of water quality, principally the reduction of nutrients and suspended solids (turbidity). A helophyte is a plant that grows in areas partly submerged in water, so that it regrows from buds below the water surface. Examples include Typha and Phragmites Australis which are typically used in constructed wetland systems used to treat wastewater. Inland shores can store and release water if needed, allowing the area to function as a climate buffer. Additional ecosystem services (e.g. recreation, fishing, aquaculture, floating infrastructure for living and working, sustainable energy production by sun, wind and water, floating agriculture, nature development) can be added depending on the

Fig. 3.2 Aerial view of Lake Ijsslemeer showing inland shore (Source: Deltares)

wishes of stakeholders. The dynamic landscape created by the Inland shore allows natural processes to function and increases the environmental quality of the area. It also serves as a carbon sink.

3.2.3   Design Criteria

The storage area was excavated and berms created in the shape of a spiral with a small pond at the centre. The design of the inland shore polder was such that a 1 m change in water level resulted in a volume of 80,000 m^3. The shape of the design allows the public to enter and experience the ecosystem (Fig. 3.2). A wind driven axial pump, which is designed to facilitate fish passage, allows water to be either taken into storage or pumped back into the lake, depending on the water levels. The pump is provided with a back-up power system. The project was designed to function as a living laboratory, with on-going monitoring allowing the designers to evaluate the impact of water level changes on water quantity and water quality, to monitor the biotic response and also to consider the issue of water safety and storage. Field tests form part of the monitoring regime, and these aim to analyse the impact of different water level regimes on water quality, water quantity, flora and fauna. In 2014, the impact of a 'natural water regime' was tested with high water levels during the winter and spring period and a natural draw down of water levels during the growing season. In 2015 a drought situation was simulated with exceptionally low water levels and in 2016 a flood event with exceptionally high water levels, was simulated.

Fig. 3.3 Inland shore with unique spiral design (Source: Deltares)

3.2.4 Key Drivers

Lake Ijsselemeer presents many challenges with respect to freshwater supply, water quality, fish stock depletion, loss of habitat and recreation facilities. In 2011, the proposed design won first prize in a competition run by one of the local water boards. This established public awareness of the project and allowed Deltares together with the Province of Noord-Holland, the regional waterboard Hoogheemraadschap Hollands Noorderkwartier, and the Government Service for Land and Water Management (DLG) to develop the concept to design/construction stage. This established the inshore concept, but also allowed for the concept of "learning by doing". This principle was key and allowed a practical on-going methodology to be developed to evaluate ecosystem services in a unique way. It also established the designers' concept that more space for water is needed in a world with a changing climate, and that water can create new opportunities for economy and quality of life in the Netherlands. Figure 3.3 illustrates the unique spiral design of the system.

3.2.5 Operation Characteristics

From 2014 to 2017 – Field tests were run, to analyse the impact of different water levels in the inshore shore lake on water quality, water quantity, flora and fauna as follows:

2013 – This was considered as a rest year allowing the vegetation to colonise and the rehabilitation of a wetland ecosystem.

2014 – This was designated as a year with a 'natural' water regime, i.e. high water levels in winter, low in summer and a natural draw down of water levels during the growing season as a function of evapotranspiration.

2015 – This was operated as a dry year. The pumping regime was set to simulate extreme low water levels.

2016 – This was operated as a wet year with the pumping regime set to simulate a flood event resulting in extremely high water levels.

3.2.6   Maintenance

The facility requires normal on-going maintenance in the form of seasonal grass cutting of banks and maintenance of the pumping system.

3.2.7   Performance

The monitoring of flora and fauna for the years 2012–2013 saw a rapid growth in vegetation particularly of submerged aquatic plants. These were mostly common land and riparian species. There was thus a pronounced increase in biodiversity. During this period there was also an increase in bird biodiversity and population size. Especially notable here, were birds of the wetland species, whose population increased after a 20 cm water level rise. There was also an increase in several rare and protected bird species. Water quality results showed a decrease in chlorophyll, ammonia, nitrates, phosphates and suspended matter. There was also an increase in water transparency in the inland shore.

Over the same period the fish population rose. Initially in 2012 small fish were observed with low population and species density. It was observed in 2013 that the fish population was mostly Roach (Rutilus Rutilus). Monitoring in 2014 observed many Pike (Esox lucius), especially of one year olds. Results for 2015 saw the monitoring of fish species typical of a fresh water habitat.

3.2.8   Resource/Recovery/Product

The amenities generated by the inland shore at Koopmanspolder illustrate the concept of ecosystem services, which are the benefits humans obtain from ecosystems. Uniquely, this project combines ecosystem services with water treatment and hydraulic infrastructure. This achieves all of the benefits associated with ecosystem services.

3.2.9 Community Involvement

The competitions, which led to the design, involved the local community. The publicity surrounding the prize, and the innovative nature of the winning concept resulted in community interest in the project being commissioned and built. The design brought the community closer to the water storage/treatment facility and also created a sense of local ownership, which would not be possible in a more traditional non nature-based "engineered" solution. Consumer acceptance is shown by the local communities' use of the wetland, as a recreation facility, and by local schools, as an open air natural history laboratory.

3.3   Case Study 2: Managed Aquifer Recharge- Llobregat River

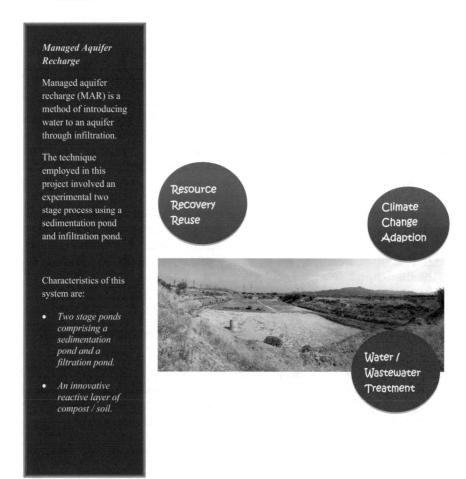

Managed Aquifer Recharge

Managed aquifer recharge (MAR) is a method of introducing water to an aquifer through infiltration.

The technique employed in this project involved an experimental two stage process using a sedimentation pond and infiltration pond.

Characteristics of this system are:

- *Two stage ponds comprising a sedimentation pond and a filtration pond.*

- *An innovative reactive layer of compost / soil.*

Resource Recovery Reuse

Climate Change Adaption

Water / Wastewater Treatment

3.3.1   Llobregat Delta Site Location

This enhancement of soil aquifer treatment (ENSAT) project was funded under the EU LIFE program 2010–2012 (www.life-ensat.eu). The project is located within the Barcelona Metropolitan Area (Area Metropolitana de Barcelona, AMB). The recharge system of Sant Vicenç dels Horts is situated in the Llobregat Delta, and is one of the pioneering zones in Spain in terms of artificial aquifer recharge. The aim of the project was to improve the quality of recharge water in the Llobregat River Delta Aquifer.

3.3.2   Managed Aquifer Recharge Concept

Managed aquifer recharge (MAR) is a technique used worldwide. It operates by introducing water to the aquifer through infiltration ponds, deep recharge wells, trenches and many other techniques (Fig. 3.4). These recharge techniques are not only used in the water scarce countries of southern Europe but they are also wide-spread in central Europe and Nordic countries. The benefits of using groundwater are that aquifers provide a store of water, which, if utilised and managed effectively, can play a vital role in:

- Poverty reduction/livelihood stability,
- Risk reduction (both economic and health),
- Increased agricultural yields resulting from reliable irrigation,
- Increased economic returns,
- Distributive equity (higher water levels mean more access for everyone),
- Reduced vulnerability (to drought, variations in precipitation).

One of the most common methods of artificial aquifer recharge is the use of infiltration ponds. Figure 3.4 shows a cross section through the reactive cell. This requires excavation of permeable terrain close to the water source. These systems also often have a sedimentation pond to improve water quality through deposition of suspended solids. Figure 3.5 shows the inlet to the system.

The main advantages of infiltration ponds compared to other recharge techniques include:

- Low construction and maintenance costs,
- Long residence time in a non-saturated zone,
- Improvement of water quality during infiltration,

The main disadvantage is the need for large areas of permeable ground.

Denitrification and reduction of organic matter in water are good examples of water treatment processes which are achieved through recharge. These processes are boosted by the microbiological activity of the non-saturated zone. They require suffi-cient residence time and the presence of easily degradable organic carbon to facilitate the growth of microorganisms. The water used for aquifer recharge in this case study

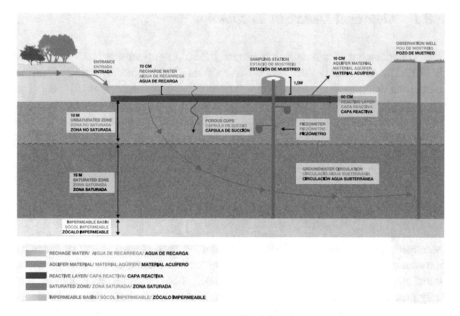

Fig. 3.4 Managed aquifer recharge (Source: ENSAT 2012)

Fig. 3.5 Inlet from Llobregat river

came from the Llobregat River, which had a low organic content (4 mg/l of dissolved carbon). The ENSAT Life + project concept was to create the natural conditions for the elimination of micro-pollutants in the water through the application of a reactive layer at the bottom of the infiltration pond. This achieves two core principles:

- Biodegradation, by increasing the organic matter available
- Adsorption, by increasing the contact area

Fig. 3.6 Inlet from sedimentation pond and piezometer well

3.3.3   Design Criteria

Water from the Llobregat river (Barcelona) is purified by percolating it through the reactive layer into the aquifer. The system of Sant Vicenç was built in 2007 and started functioning between 2008 and 2009, under the management of the Catalan Water Agency (ACA). The system includes a sedimentation pond of 4000 m² and an infiltration pond of 5600 m² The extraction rate is of the order of 1.0 m³/m² surface area/day. The design flow was 250 m³/h. The operative flowrate was in the range 200–500 m³/h. The system was designed to receive water from the Llobregat River via the deviation of Molins de Rei, and reused water from El Prat del Llobregat water reclamation plant (Fig. 3.6).

The design requirements for the reactive layer were to provide a constant supply of dissolved organic carbon, be readily available, (to ensure the project could be replicated anywhere), be of low cost (reuse of sub product or waste from another process), be safe in terms of human health and easy to handle with regard to installation.

A compost was chosen, i.e. a natural product obtained from the crushing of garden waste. This waste was mixed with local soil material to ensure maximum infiltration. A small amount of highly adsorbent material was finally added to the mix (clay and iron oxide) to boost these processes. Over 1500 m³ of compost mixed with the local soil was spread at the bottom of the pond to a depth of 1 m. Figure 3.7 shows the two connected ponds.

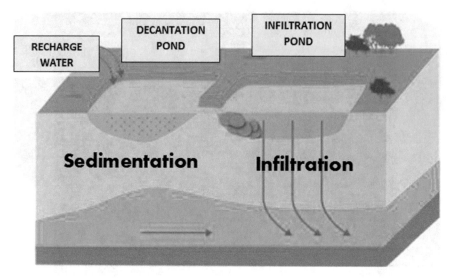

Fig. 3.7 Two connected sedimentation and filtration ponds

3.3.4 Key Drivers

The key drivers in this project were:

- Assessment of the feasibility of utilising an NBS comprising a substrate layer to enhance the treatment of low quality water prior to infiltration into the aquifer
- Development of a series of modelling tools to model the processes within the system
- Dissemination of results throughout Europe

The project was also part of a wider reuse scheme where El Prat de Llobregat water reclamation plant, located upstream treats water to a basic standard which is then discharged through the aquifer recharge scheme to provide a buffer against water shortages, within the aquifer.

3.3.5 Operation Characteristics

The dimensions of the pond of Sant Vicenç enabled experimental work to be carried out under real conditions and with a high level of control of the system. The system installed 16 control points where piezometric levels were measured both outside and inside the pond. In addition, during the project a third measurement point inside the pond was added to the existing control equipment. A piezometer is an

Fig. 3.8 Sedimentation pond

instrument for measuring pressure or compressibility; especially for measuring the change of pressure of a material subjected to hydrostatic pressure. Some of the piezometers installed were environmental piezometers, installed to obtain samples of the entire profile. Some other piezometers were multilevel piezometers, located at specific depths to analyse the variation in water quality at different depths. To measure the chemical parameters during the recharge process, porous cups were buried under the pond at 1, 2 and 5 m depth. These capsules enabled the gathering of water samples in the non-saturated zone. Other devices, such as tensiometers and humidity sensors, were also placed to measure the saturation conditions of the terrain. Figure 3.8 shows a photo of the sedimentation pond.

3.3.6 Reactive Compost Layer Performance

The final design of the reactive layer is composed of 50% of vegetal compost coming from a compost plant near to Barcelona and 50% of local sand and gravels from the bottom and the slopes of the infiltration pond. This mixture was enriched with 1% of red clay and less than 0.1% of iron oxides. An extra layer of 5 cm of local sand and gravel was put at the bottom to reduce floating elements. The ENSAT project proved the efficiency of the reactive layer in terms of reduction of emerging pollutants, such as gemfibrozil and carbamazepine epoxy in recharge water. It also improved the quality of the aquifer water. The processes which promote pollutant removal are the increased amount of dissolved organic carbon available and the increased level of adsorption supplied by the compost. The impact of the reactive layer is local, only on the non-saturated zone, while the good quality water will remain available for the future in the aquifer.

3.3.7 Resource/Recovery/Product

The utilisation of a reactive compost layer led to the removal of organic pollutants and to a higher quality recharge water. This process provides an alternative solution to conventional pre-treatment needed for aquifer recharge, which requires more energy and reagents. The prerequisites in designing a reactive layer are the availability of the selected materials and their costs. To implement a reactive layer in a new site, part of the methodology established in the ENSAT project can be applied.

Further Information
   http://www.amb.cat
   http://www.life-ensat.eu/

3.4   Case Study 3: Green Port Concept: Rotterdam Mallegats Park

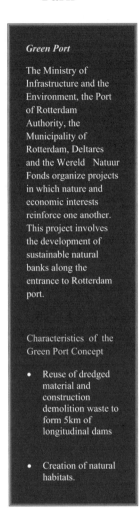

Green Port

The Ministry of Infrastructure and the Environment, the Port of Rotterdam Authority, the Municipality of Rotterdam, Deltares and the Wereld   Natuur Fonds organize projects in which nature and economic interests reinforce one another. This project involves the development of sustainable natural banks along the entrance to Rotterdam port.

Characteristics of the Green Port Concept

• Reuse of dredged material and construction demolition waste to form 5km of longitudinal dams

• Creation of natural habitats.

3.4.1   Mallegats Park, Rotterdam

Deltares, the Ministry of Infrastructure and the Environment, the Port of Rotterdam Authority, the Municipality of Rotterdam and the Wereld Natuur Fonds (the Dutch branch of the World Wide Fund for Nature) organise projects in which nature and economic interests reinforce one another. The first project was officially launched on 2 October 2013. The development of sustainable natural banks along the river is an integral component of the *"Green Port"* concept in the region. The site of this case study is part of this strategy and it is located in Mallegats Park on the southern side of the port of Rotterdam.

3.4.2   Green Port Concept

The Green Port nature-based design project has two important functions. The first is to capture, naturally, the tidal sediment/sludge. This is done using dams, to capture via sedimentation the suspended particles in the river system. The second is to use groynes (A groyne is a rigid hydraulic structure built from an ocean shore (in coastal engineering) or from a bank (in rivers) that interrupts water flow and limits the movement of sediment) to reduce wave velocity allowing the restoration of the natural shore. Groynes are hydraulic structures built at right angles to the shore. Sediment from longshore drift is trapped behind the groyne to form sandbanks. These natural processes are augmented by dredged material from the harbor being recycled to promote sand bank formation. The site will increase water quality, provide natural value and increase citizens' awareness of the tidal system they live in. This natural shore will then provide ecosystem services in the form of recreation facilities such as cycling, fishing and nature walkways. Figure 3.9 illustrates the creation of natural habitats.

3.4.3   Design Criteria

This project utilises NBSs to solve a marine engineering problem and to create added value through the creation of public recreation facilities/ecosystem services. The marine engineering is focused on the use of tidal dams and groynes to accumulate sediment, keeping the shipping channel free from sediment and creating sand banks to reduce tidal erosion. The dredge material removed from the shipping channel is also recycled, increasing the accumulation of solids and promoting sank bank formation. The recreation strand takes as its focus the 'river as a tidal park', and by utilising the marine engineering innovations to create cycle paths, walkways and fishing areas, aims to give the general public a sense of the river as a diurnal amenity.

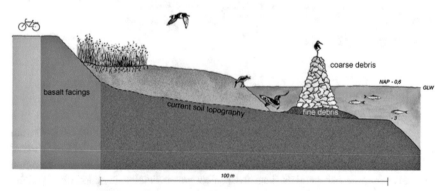

GHW = average high tide   GLW = average low tide

Fig. 3.9 The green port concept showing the creation of natural habitats to protect the shoreline (Source: www.deltares.nl)

The project consists of constructed dams to capture solids. The dams, allow the accumulation of suspended materials behind the dam walls. This traps the suspended materials and prevents them settling in and blocking the main shipping channel. The dams and groynes also serve as a site to recycle the dredge material from the shipping channels. Figure 3.10 shows an example of the groynes.

Characteristics of the project are as follows:

Five kilometres of natural shoreline banks
Along the banks of the Landtong, from the ferry slip to the Maeslantkering, longitudinal dams were constructed in the water (Fig. 3.10). These are stone dams which run parallel to the banks. The dams protect the banks from ship-induced waves. The harbour's waste material is used to create dams. Material from old quay walls, as well as support and concrete piles, were stacked to form a dam through which water could flow. eas The water depth between the dam and the banks will also be reduced. Waste material from concrete constructions in the surrounding area will be used to make this area shallower.

Mud flats and groynes
Between the groynes an intertidal area is formed. This transitional zone from water to land dries up when the tide goes out and floods when the tide comes in. In this 'mud flat' habitat, salt-loving plants feel at home as shown in Fig. 3.11. An intertidal area offers a peaceful, living and foraging environment for various fish and bird species. This nutrient-rich habitat is an ideal environment for young fish and shrimp. Migratory birds and migratory fish, such as the eel and the salmon, rest in this habitat. It is hoped that if the Atlantic Sturgeon survives, it will inhabit the site, as this is a suitable habitat for this species.

Fig. 3.10 5 km of natural shoreline banks

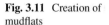

Fig. 3.11 Creation of mudflats

3.4.4   Key Drivers

Rotterdam port, one of the busiest in the world has historically struggled to keep the shipping channels at the required depth to allow access to all shipping. Sludge accumulation and erosion due to wave motion have resulted in a constant need to dredge areas of the harbour prone to silting. This resulted in costly transport, treatment and disposal processes. Innovative marine engineering processes, allow these practices to be reduced as the dredge material is now used on site in the dams and at the groynes.

3.4.5   Operation Characteristics

The accumulation of sediment within the dam structure and behind the groynes, shown below, results in the creation of sandbanks. These serve a twofold purpose, in that they reduce the erosion effects of wave motion, and they will be colonised by grasses. It will also lead to the creation of a recreational area for walking, biking and fishing.

3.4.6   Performance

The mechanisms involved are primarily sedimentation and settlement. The dams and the groynes slow down the motion of the water allowing the water to shed its load. This load accumulates within the wall of the dam and at the groyne wall. Figure 3.12 illustrates the concept.

Further Information

http://www.rotterdam.nl/getijdenpark
http://www.landtongrozenburg.nl/green-port.html?l=2
www.deltares.nl

Fig. 3.12   Green port concept

3.5   Case Study 4: Restoration of Ecosystems: Granollers, Barcelona

Espai Natural de Can Cabanyes

Can Cabanyes is a highly industrialised area near the river Congost in the municipality of Granollers, Barcelona.

Characteristics of this system are:

- *Tertiary Treatment of industrial wastewater*

- *Restoration of river water quality*

- *Provision of ecosystem services*

- *Education facilities*

	Site Location:	Project Partners
UNIVERSITAT POLITÈCNICA DE CATALUNYA BARCELONATECH. Group of Environmental Engineering and Microbiology **Universitat Politècnica de Catalunya BarcelonaTech. Environmental Engineering and Microbiology Research Group (GEMMA-UPC), Spain.** www.gemma.upc.edu	Can Cabanyes, Granollers Barcelona Region, Spain	GEMMA-UPC Granollers City Council

3.5.1   Site Location Can Cabanyes

The Surface Flow Constructed Wetland (SFCW) was created in the peri-urban park of Can Cabanyes (Granollers, Catalonia, northeastern Spain), located next to a highway, an old landfill, a large conventional urban WWTP, a solid waste treatment plant and a frequently used racetrack (Fig. 3.13). The wetland has a Mediterranean coastal climate, characterised by warm temperatures, mild winters (average minimum temperatures above 5 °C) and hot and dry summers (average temperatures of around 25 °C, particularly in July). The average annual precipitation is 647 mm. The intensity and frequency of rainfall events vary throughout the year.

3.5.2   Concept

Can Cabanyes is situated in a peri-urban industrialised area near the river Congost in the municipality of Granollers (Barcelona). This project aimed to clean up and restore this river environment with a series of measures, which can reconcile environmental improvement with the use of the area (ecosystem services). One of the measures was to build a 1 ha constructed wetland, which is fed with the effluent from the Granollers WWTP. The final reclaimed water from the natural treatment system is reused for urban and agricultural purposes (i.e. street cleaning and irrigation of public parks). As well as conventional water contaminants (i.e. Total suspended solids (TSS), chemical oxygen demand (COD), ammonium and faecal microbial indicators), this project also focused on the efficiency of the system in removing pharmaceuticals and personal care products (PPCPs). The selected compounds include a large range of chemicals commonly used by humans (analgesics, anti epileptics, anti lipidics and fragrances).

The focus of the project was to develop an NBS, which could achieve the required technical treatment efficiency and also provide added value through ecosystem services with the site serving as a recreational area for the locality and environmental education facility. Figure 3.14 shows the educational communication facility at the site.

3.5.3   Design Criteria

The constructed wetlands, consisted of a zone planted with Phragmites australis; a zone planted with Typha latifolia; a deep zone free of macrophytes and an island. The system provides treated water for park irrigation and several ecosystem services. The flow to the wetland was calculated based on an influent ammonium concentration of 30 mg N/L, and the objective was to produce an effluent with an ammonium concentration lower than 2 mg N/L. The SF CW is a single cell system

Fig. 3.13 Industrial facility adjacent to Can Cabanyes

Fig. 3.14 Added value
through ecosystem services

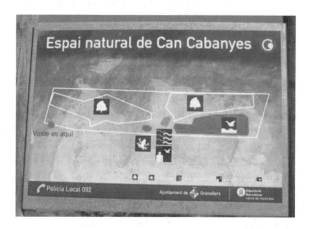

with an elongated shape and a surface area of 1 ha (maximum length and width of around 189 m and 53 m, respectively). It was planted with 2250 transplanted units of Phragmites australis and Typha latifolia, creating different vegetation zones and increasing the ecological variability of the system.

3.5.4   Key Drivers

The final reclaimed water from the natural treatment system is reused for urban and agricultural purposes (i.e. street cleaning and irrigation of public parks).

3.5.5   Operation Characteristics

A number of published studies have evaluated the scientific and economic performance of the system (Llorens et al. 2009; Alfranca et al. 2011; Matamoros et al. 2008). The wetland began operating in April 2003. The vegetation grew very quickly and by September 2003 the planted surface was dense. Between the initiation of the constructed wetland and April 2006, the wetland received approximately 100 m³/day of secondary effluent from the Granollers WWTP, operating with a hydraulic retention time (HRT) of around 1 month and a hydraulic loading rate (HLR) of 1 cm/day. The Granollers WWTP serves an equivalent population of approximately 154,000 people and treats urban wastewater (55%) mixed with a considerable amount of industrial discharge (45%), without complete nitrification. The subsurface flow constructed wetland (SF CW) influent only represented 0.4% of the WWTP flow rate. Because the plan is to reuse the effluent from the wetland, in April 2006 the flow to the wetland was increased to approximately 250 m³/day in order to check the operation efficiency of the wetland under the new conditions. As a result, the HRT decreased to 12.4 days and the HLR increased to 2.5 cm/day. In both periods, the effluent from the SF CW was discharged into the Congost River, as is the case for the other treated effluents from the Granollers WWTP.

3.5.6   Performance

Table 3.1 summarises the principal removal mechanisms in the system. The low COD and TSS removal rates observed in the SF CW of the peri-urban park of Can Cabanyes are linked to the low incoming pollutant concentrate ions and to its strong eutrophic character, which is in fact a consequence of the WWTP effluent properties (i.e. high ammonium concentrations). However, the created wetland efficiently removed ammonium (80%), faecal bacteria indicators (around 2 logarithmic units of Faecal Coliforms) and the amount of PPCPs discharged into the system, resulting in a quite good quality effluent. A seasonal pattern was clearly observed in ammonium concentrations, which shows the temperature dependence of the mechanisms involved in ammonium removal.

3.5.7   Resource/Recovery/Product

Effluents from the SF CW were sampled between 2003 and 2006 for physical and chemical parameters and faecal bacteria indicators. In addition, 8 PPCPs were measured in June 2005 and February 2006. The system showed a good reliability for

Table 3.1 Removal mechanisms Can Cabanyes wetland system

Wastewater constituent	Removal mechanism
Gross solids	Coarse/fine screening
	Fats oils and grease removal(flotation)
	Grit removal
Suspended solids	Sedimentation
	Filtration
Soluble organics	Aerobic microbial degradation
	Anaerobic microbial degradation
Nitrogen	Ammonification
	Nitrification
	Denitrification
Phosphorus	Precipitation/organic removal
Pathogens	Sedimentation
	Filtration
	Natural die-off
	Predation
	UV irradiation

ammonium and faecal bacteria removal, with average ammonium efficiencies between 64% and 87% and a removal of approximately 2 logarithmic units of Faecal Coliforms. The results for PPCPs demonstrated that the wetland has a good capacity for removing a large variety of these compounds; the removal efficiencies were higher than 70% for most of them, with the exception of clofibricacid (34%) and carbamazepine (39%), (Alfranca et al. 2011). **The analysis of the physical and chemical parameters in the present study revealed that the SF CW effluent could be suitable for agricultural irrigation.**

The externalities of the wetland were evaluated using the travel cost method. The travel cost method is widely used to estimate the value of natural resources, particularly in recreational sites (Gurluk and Rehber 2008). The value of the wetland is expressed in terms of the price of the water that flows through the system, which is estimated to range from 0.71 to 0.75 €/m³. The value of positive externalities (1.25 €/m³) was greater than private costs (from 0.50 to 0.54 €/m³). These results constitute empirical evidence that created wetlands in peri-urban parks can be considered to be a source of positive externalities when used in environmental restoration projects focusing on the reuse of treated wastewater. This study also illustrates the small influence of the hydraulic infrastructure depreciation costs on the overall cost of constructed wetlands (less than 10%), and the low investment costs of constructed wetlands in comparison with operation and maintenance costs (less than 10% of total private costs).

3.6   Case Study 5: Constructed Wetlands Sludge Treatment Systems: La Guixa, Barcelona Region

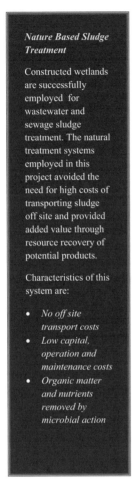

Nature Based Sludge Treatment

Constructed wetlands are successfully employed for wastewater and sewage sludge treatment. The natural treatment systems employed in this project avoided the need for high costs of transporting sludge off site and provided added value through resource recovery of potential products.

Characteristics of this system are:

- *No off site transport costs*
- *Low capital, operation and maintenance costs*
- *Organic matter and nutrients removed by microbial action*

	Site Location:	Project Partners
 **Universitat Politècnica de Catalunya BarcelonaTech. Environmental Engineering and Microbiology Research Group (GEMMA-UPC), Spain.** www.gemma.upc.edu ivet.ferrer@upc.edu	Estacio depuradora 'aigues residuals (EDAR) de Santa Eulalia de Riuprimar. EDAR Les Cases Noves, Les Masies de Roda. La Guixa. Barcelona Region, Spain	Spanish Ministry of Environment Depuradores d'Osona S.L. The Technical University of Catalonia. (MMARM, Project 087/PC08).

3.6.1   La Guixa Site Location

La Guixa is a small WWTP (1000 Population Equivalent, PE) located in the province of Barcelona (Spain), which treats 100 m^3/d of urban wastewater in an activated sludge process with an extended aeration system. Five wetlands with a total surface of 210 m^2 were established in 2007 to treat waste activated sludge.

3.6.2   Sludge Treatment Wetlands Concept

Sludge treatment wetlands (STW) consist of shallow tanks (beds) filled with a gravel layer and planted with emergent rooted wetland plants. These beds provide a drying phase and a mineralisation phase as in conventional reed bed systems. In these systems, secondary sludge is usually pumped and spread on the wetland's surface. The sludge fed is rapidly distributed over the wetland and part of its water content is rapidly drained by gravity through the gravel layer; while another part is evapo-transpired by plants. In this way, a concentrated sludge residue remains on the surface of the bed. After some days without feeding (resting time), thickened sludge is spread on the surface once again, starting the next feeding cycle.

3.6.3   Design Criteria

The STW technology is an environmentally regenerative and sustainable technique, which does not require any chemical additives and sophisticated electronic control. The technology has extremely low operating, maintenance and energy costs. The technological principles are based on biological, chemical and physical processes that occur in natural wetlands. Depuradores d'Osona S.L. constructed nine STWs at their rural WWTPs.

Site selection criteria were based on an assessment of the following:

• That there was enough sludge being produced to warrant on-site treatment,
• There was sufficient land available to construct the required area of wetland,
• There was sufficient distance between the proposed STW and the central treatment area in Vic to prove cost effective in terms of reducing transport costs,
• The WWTP was sufficiently automated to allow for the modifications necessary to run the STW.

La Guixa is one example of this strategy and comprises a WWTP (1000 PE) located in the municipality of Vic, comarca Osona, Catalyuna (Spain), which treats urban wastewater in an activated sludge system with extended aeration. Average annual rainfall is of the order of 650 mm while temperatures range between 8 and 28 °C. In 2007, 5 basins with a total surface of 210 m^2 were constructed with concrete walls and planted with P. australis. Design criteria are described in Table 3.2.

Table 3.2 Design characteristics of La Guixa sludge treatment system

PE	1000
Sludge loading rate (kg dry matter/m²/year)	50
Average influent Total Suspended Solids (%)	0.4
Total surface area (m²)	210
Number of beds	5

Uggetti et al. (2012b)

3.6.4   Key Drivers

The existence of small and remote communites (<2000 PE) in the province of Catalyuna led to the practice of tankering sludge from each treatment plant to a central plant for treatment. The construction of STWs on-site remove this cost. Depuradores d'Osona S.L. have twenty seven.

WWTPs of various sizes within their jurisdiction. The smaller plants were not constructed with any system of sludge drying or dewatering. Therefore, the liquid sludge was stored to reduce volume before being transported to the central wastewater treatment plant at Vic for treatment. Here, the sludge was mixed with sludge from the Vic WWTP (Fig. 3.15) dewatered, and tankered to Landfill. STW facilitated the following:

- Sludge treatment at source (i.e. where it was generated),
- Reduced operational costs,
- Reduced transport costs and other associated risks,
- Improved sludge management practices and the environmental impact of sludge.

Operation Characteristics

Figure 3.16 illustrates the operation of the system. The sludge is distributed over the wetland and part of its water content is rapidly drained by gravity through the gravel layer; while another part is evapotranspired by plants. In this way, a concentrated sludge residue remains on the surface of the bed where, after some days without feeding (resting time), fresh thickened sludge is spread, starting the next feeding cycle.

During feeding periods, the sludge layer height increases at a rate of around 10 cm/year. When the layer approaches the top of the banks or walls surrounding the STW (usually after 8–12 years), feeding is stopped. The sludge remains in the beds for a final resting period (from 1–2 months to 1 year), aimed at improving sludge dryness, mineralisation and dewatering. This resting period improves the sludge dryness and mineralisation. The final product is subsequently withdrawn Fig. 3.16. Biosolids obtained from the treatment are suitable for agricultural uses. Figure 3.17 shows the influent loading system.

Essentially, the STW is a vertical flow reed bed, and the reeds provide aerial and subterranean stalks and roots. These supply drainage through windrock and also through root penetration increasing the amount of hydraulic pathways through the sand and gravel layers. It is estimated that about 1% of the sludge liquor is absorbed

Fig. 3.15 Activated sludge wastewater treatment plant

Fig. 3.16 Sludge biomass

by the roots, 9% is lost through evapotranspiration and the remaining 90% is filtered to the draining layers. This liquid is then recycled back to the treatment plant. Table 3.3 shows the operational design characteristics for one of the systems.

3.6.5 Performance

The waste activated sludge generation at La Guixa was 500 m³/year. Sludge production in the STW was 66 m³/year. Pump electricity consumption was 50 kWh/year. CO_2 emission rate was 0.25 kg/m².d (Uggetti et al. 2012a, b).

Fig. 3.17 Influent pumped
onto wetland system

Table 3.3 La Guixa design
characteristics

Resting period between feeding	8
Bed depth (m)	1.7
Gravel volume per system (m³)	21
Sludge storage capacity per system (m³)	42
Average operating cycle (years)	10

Uggetti et al. (2012a, b)

Fig. 3.18 End product
showing mineralised
sludge

3.6.6   Resource Recovery Product

The construction of STWs are economically viable, the main cost being the construction of the reed bed. In the case of La Guixa and other plants in the area, investment has been recovered in 6–8 years. Operational costs are a minimum. The STW has reduced labour costs in the central WWTP. No chemical input was required (Fig. 3.18).

3.7   Case Study 6: Mill Pond

Mill Pond

Located in the municipalities of Gavà, Viladecans and SantCliment de Llobregat (Barcelona), Spain.

The mill pond acts as a retention basin, storing rainwater during storm events. There are two tributaries of the Sant Lloerenc which also enter the Mill pond, during storm events.

Characteristics of this system are:

- Provides flood protection,
- Reduces risk disaster
- Mitigates the effects of climate change

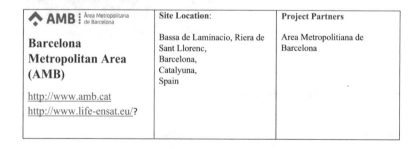

⌂ **AMB** ⋮ Àrea Metropolitana de Barcelona	Site Location:	Project Partners
Barcelona Metropolitan Area (AMB) http://www.amb.cat http://www.life-ensat.eu/?	Bassa de Laminacio, Riera de Sant Llorenc, Barcelona, Catalyuna, Spain	Area Metropolitiana de Barcelona

3.7.1   Site Location

The project is located in Bassa de Laminacio, Riera de Sant Llorenc, Catalyuna. It is a mill pond constructed to reduce the risk of flooding from the San Llorenc river in the municipalities of Gavà, Viladecans and Sant Climent de Llobregat (Barcelona).

3.7.2   Concept

A millpond is a pond constructed to store storm water through a storm event. The stored water is later released when the flood event has passed, and the water from the mill pond can be safely released. The mill pond provides flood protection, reduces risk disaster and mitigates the effects of climate change. The characteristics of a site which make it suitable for an online storage solution, such as a mill pond, can be summarised as follows:

- A suitable location within the catchment for the purpose intended with sufficient storage volume.
- A suitable site for the impoundment structure – for example taking advantage of a narrower part of the valley to allow the dam to be shorter.
- A wide floodplain that allows a low dam height to be deployed.
- A relatively impermeable foundation.
- Suitable foundation conditions for supporting the dam and control structures.
- Suitable access for construction, operation and maintenance.
- The availability of suitable construction materials on or near the site.
- Minimum adverse impacts on landowners, land-use and local residents.
- Minimum adverse impacts on the environment.
- Opportunities for environmental enhancement (Fig. 3.19).

3.7.3   Design Criteria

The mill pond was sized at 157,480 m³. This was the design volume to deal with a 100-year storm event with an inflow of 20 m³/s into the mill pond through the inlet by-pass structures. Figure 6.1 illustrates the operation of the storage facility. The inflow hydrograph (shown in blue) starts to overflow into the mill pond structure at flowrates in excess of 20 m³/s. The flood storage pond provides storage during peak flood events and slowly releases the water into the catchment downstream after the storm event has subsided. The retention of this volume of water prevents flooding of the towns of Viladecans and Gava.

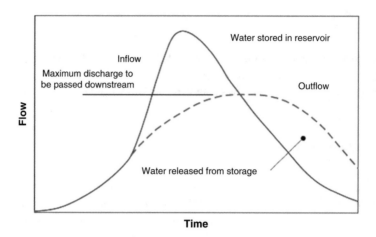

Fig. 3.19 Theoretical operation of flood control storage

3.7.4 Key Drivers

The Sant Llorenc river flooded regularly and caused extensive damage to an area in the municipalities of Gavà, Viladecans and Sant Climent de Llobregat (Barcelona). A solution to this recurring problem was the driver behind the project.

3.7.5 Operation Characteristics

The storm water from three tributaries is gathered in the mill pond. The construction and operation of the nature-based flood storage solution are detailed in the following photos (Figs. 3.20, 3.21, and 3.22).

Fig. 3.20 Collector bypass

Fig. 3.21 Inlet structure

Fig. 3.22 Storage reservoir

3.8   Case Study 7: Nature-Based Nutrient Recovery

Nutrient Recovery

The Netherlands is characterised by intensive agricultureresulting in high concentrations of nutrients and pesticides, leading to eutrophication,and cyanobacteria blooms, in some areas. There is little space for extensive solutions.

Characteristics of nature-based nitrate removal

• Denitrification using microorganisms in soil and groundwater using an energy source comprising wood chips, ethanol, etc.

Characteristics of nature-based phosphate removal:

Chemical immobilisation using iron coated sand and a purifying bench.

Restoring Ecosystems

Resource Recovery Reuse

Water / Wastewater Treatment

| Deltares
Enabling Delta Life
Deltares
The Netherlands
www.deltares.nl | **Site Location**:

Flower/bulb growing regions of north Holland and south Holland in the Netherlands | **Project Partners**

The project partners are the Hoogheemraadschap van Rijnland, which is the oldest water board in the Netherlands, and three Dutch research institutes Alterra, Deltares and Arcadis. |
| --- | --- | --- |

3.8.1   Site Location

The "nutrient removal from tile drainage" project is located in the flower/bulb growing regions of north and south Holland in the Netherlands.

3.8.2   Concept

The project is an application of low cost robust water technology. Nutrient removal occurs within the drainage cycle. The technology seeks to remove nitrates, by utilising wood chips and ethanol as sources of carbon, to stimulate microorganism activity in the soil and groundwater. This promotes denitrification, which converts the nitrate to nitrogen gas. The technology seeks to remove phosphates by immobilising them and removing them from the drainage cycle. A binding material, e.g. iron coated sand, is added to the soil surrounding the pipe drain to remove phosphates. Electrocoagulation, at the end of the drainage channel, and a purifying bed, comprising of two materials, iron coated sand and poly-aluminium chloride/sand were also part of the designs used in the studies.

3.8.3   Design Criteria

Nitrates
In designing for nitrate removal two methods were employed. The first required excavating the tile drains and surrounding the pipe drain with a sand/soil which contained wood chips and in some examples beet pulp (Figs. 3.23 and 3.24). As the irrigation water drained from the pipe, it came into contact with this source of carbon which promoted denitrification.

Fig. 3.23 Nitrate removal
from tile drains

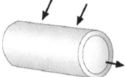

Fig. 3.24 Sand/soil mixture with wood chips

The other method employed a methanol dosing reactor placed at the discharge point of the drain, which allowed the drain water to pass through it, thus promoting denitrification and the release of nitrogen gas.

Phosphates

In designing for phosphate removal a number of methods were evaluated. The first excavated the drains and filled the pipe surround with iron coated sand. The second method was referred to as a phosphate removing reactor which promoted electrocoagulation of the P at the end of the drainage channel. The third method, known as a purifying bench, consisted of two materials, iron coated sand and poly-aluminium chloride/sand. This allowed the drainage water to pass through and the phosphate to be bound.

3.8.4   Key Drivers

The high pressure on land for agriculture and human settlement has resulted in intensive agricultural and horticultural practices within the Netherlands. Agriculture and horticulture account for 10% of the Dutch economy. This has resulted in intensive crop cultivation with high inputs of the nutrients, nitrate and phosphates and also of pesticides. The intensive drainage systems that have developed with these practices, tile drains or pipe drains, carry a water rich in nutrients that can promote eutrophication and give rise to cyanobacterial blooms. This has been a recurrent problem over a number of years. Given the high density of population and agriculture there is little space for extensive solutions.

3.8.5   Operation Characteristics

The additives to the soil surrounding the pipe drains, i.e. the wood chips and the iron coated sand required excavation of the drains and backfill of the material surrounding the pipe drains.

The ethanol reactor, the purifying bench and the electrocoagulator were all installed to allow passage of the drainage water through each one (Fig. 3.25).

3.8.6   Performance

Nitrate removal
Denitrification is effected by denitrifying microorganisms which reduce nitrates to nitrites and then to nitrogen gas. The addition of wood chips and ethanol promotes bacterial activity by supplying a source of carbon. Nitrate levels were reduced from a reference level in the range 5–7 mg N/l to zero using wood chips and beet pulp as the surrounding pipe media. The ethanol reactor showed removal rates of between 2% and 100%. The 100% removal rate was observed after a period of 8 weeks in operation.

Phosphate removal
Phosphate is precipitated using the metals aluminium or iron. All methods of removal achieved phosphate removal rates of between 80–95%. The removal rates for the pipe drains enveloped in iron coated sand were 94%. The puri bench saw Ortho-P removal rates of 80% and Total P removal rates of 90%. Hydraulic permeability varies through the systems with the iron coated gravel more permeable than the aluminium sand. This is important in terms of capacity and sizing.

Fig. 3.25 Purifying bench
for phosphate removal

3.8.7   *Resource/Recovery/Product*

The projects quoted focused on removal of nutrients. However, there is scope to evaluate the potential to use the technologies to adapt a process, which uses the removal material as a product in another process.

Further Information
www.deltares.nl

3.9   Case Study 8: Nature-Based Wastewater Treatment Systems in Slovenia

Nature-Based Wastewater Treatment, Slovenia

Constructed Wetlands (CW) are successfully employed as an alternative technology

for wastewater and sewage sludge treatment for rural communities in Slovenia.

Characteristics of the Limnos CW systems:

- *Easy to set up and maintain*

- *Low investment costs*

- *Low operation and maintenance costs*

- *Organic matter and nutrients removed by microbial action*

Restoring Ecosystems

Water / Wastewater Treatment

Resource Recovery Reuse

LIMNOS Ltd

http://www.limnos.si/eng/index.php

	Site Location:	Project Partners
LIMNOS **LIMNOS Ltd** http://www.limnos.si/eng/index.php	Sveti Tomaz (500 pe) Sodinci (1,100 pe) RIBOGOJSTVO Goricar Fish Processing factory, Slivje, Slovenia.	Limnos Ltd. was established in 1994. The company is engaged in research development and application of NBS in Slovenia, Croatia, Montenegro Italy, Macedonia, Albania, Bosnia & Herzegovina.

3.9.1   Site Location

NBS applications in Slovenia are presented at a number of different locations:

- The village of Sveti Tomaz located in South East Slovenia. This project comprised a total wastewater treatment solution for a small village of 500 PE.
- The village of Sodinci, comprising a constructed wetland wastewater treatment system for a PE of 1100.
- An industrial application of a natural wastewater treatment system located in a fish processing factory, Ribogojstvo Goricar near Slivje.

3.9.2   Concept

Horizontal Subsurface Flow Treatment Systems
Horizontal subsurface flow (HSF) wetlands typically employ a gravel bed planted with wetland vegetation. The water is kept below the surface of the bed and flows horizontally from the inlet at one end through a flat gently sloping bed of 1–2% slope to the outlet. The water level in the bed is typically controlled by an outlet level control device.

The basic schematic of the LIMNOWET^R system, which is patented, is as shown in Fig. 3.26. The treatment stages are:

- Primary treatment comprising inlet coarse screens, sedimentation tank,
- Secondary treatment comprising a filtration bed followed by a treatment bed and a polishing bed.

The actual treatment system designed and installed may differ from the concept depending on the site specific requirements of the project.

3.9.3   Application 1: Sveti Tomas Total Wastewater Treatment System

The function of the reed bed system in Sv. Tomaz is to provide a total wastewater solution for a small community. The reed bed system includes a STW which deals specifically with the sludge in the wastewater. This STW allows for onsite treatment of the sludge and eliminates the need for off-site tankering, thereby significantly reducing operational costs for the community. The effluent from the STW is subsequently treated by the reed bed system (Fig. 3.27).

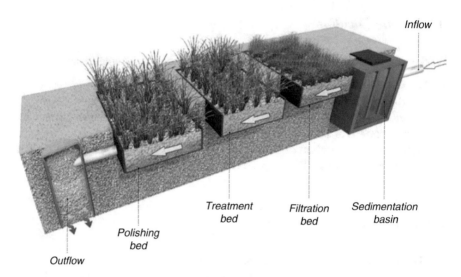

Inflow

Treatment bed

Filtration bed

Sedimentation basin

Polishing bed

Outflow

Fig. 3.26 Schematic showing LIMNOWET^R constructed wetland operation (Source: Limnos)

Fig. 3.27 Sveti Tomaz treatment system showing LIMNOS four stage treatment system with sludge treatment

3.9.4 Design Criteria

The following design was proposed and built for a PE of 500. A sedimentation tank of size 50 m^3 was installed, followed by a screen which allowed the reed beds to be fed by gravity. The reed beds had a total sizing of 1250 m^2. This consisted of three horizontal beds. Bed 1 was sized at 12 m × 25 m (total area 300 m^2), Bed 2 at 25 m × 25 m (total area 625 m^2) and Bed 3 at 13 m × 25 m (total area 325 m^2). Figure 3.28 shows the inlet channel to the horizontal reed bed.

Fig. 3.28 Inlet channel to
horizontal reed bed

Fig. 3.29 Coarse inlet
screens

The first bed functioned as a filtration bed, the second as a treatment bed and the third bed as a polishing bed. The effluent is discharged to a nearby watercourse. The beds are all lined with an impermeable liner, and the media is a washed gravel of single grading (Fig. 3.28). The beds are planted with Phragmites Communis (Fig. 3.29).

A separate STW was constructed adjacent to the sedimentation tank to treat the sludge (Fig. 3.30). This was sized at 12 m × 12 m Total area (144 m²). It is lined and has a layer of washed gravel and is planted with Phragmites Communis. The drained liquid from the STW enters the filtration reed bed for treatment.

Fig. 3.30 Sludge reed bed

3.9.5   Key Drivers

The local community/municipality can operate and maintain the community wastewater treatment system without necessary external inputs that could include chemicals, skilled labour, technicians, plant and machinery. It also facilitates the treatment and management of municipal sludge on site.

3.9.6   Operation Characteristics

The plant is located below the village of Sv. Tomaz and wastewater flows by gravity to the sedimentation basin. This allows for solids accumulation and also allows the supernatant to overflow, through screens, onto the reed bed system. Pipework is used to distribute the influent onto the treatment reed bed. It proceeds from this bed, by plug flow, to the treatment bed and then to the polishing bed. The accumulated sludge in the sedimentation basin is pumped monthly to the STW, where it is dewatered and stored. The dewatered solids are mineralised, while the filtrate water is passed to the reed bed.

3.9.7   Application 2: Sodinci Wastewater Treatment System

The village of Sodinci, in south east Slovenia, has a PE of 1100. A reed bed facility was constructed to treat the village wastewater. This was a Horizontal-flow reed bed as shown in Fig. 3.31.

Fig. 3.31 Constructed
wetland system for village
of Sodinci

Fig. 3.32 Treatment of wastewater from fish processing factory, Slovenia

3.9.8   Application 3: Industrial Site, Ribojstvo Goricar

This application is the treatment of wastewater from a fish processing factory special-
ising in the production of rainbow trout for the domestic Slovenian market. Figure 3.32
shows the fish processing unit. The wastewater from the factory is treated in a horizontal-flow reed bed, which discharges to a local water course (Figs. 3.32 and 3.33).

3.9.9   Results

Table 3.4 summarises the treatment and removal mechanisms within the systems in
Slovenia.

   Table 3.5 presents results for the villages of Sveti Thomas and Sodinci. Both systems achieved removal efficiencies for Chemical Oxygen Demand (COD) in excess
of 93% and for Biochemical Oxygen Demand (BOD) in excess of 97%. Results for
the Industrial Fish Site, Ribojstvo Goricar were not available from the Local Authority.

Fig. 3.33 Natural wastewater treatment system, fish processing factory, Slovenia

Table 3.4 Removal mechanisms in Slovenia natural wastewater treatment systems

Wastewater constituent	Removal mechanism
Gross solids	Coarse/fine screening Fats oils and grease removal(flotation) Grit removal
Suspended solids	Sedimentation Filtration
Soluble organics	Aerobic microbial degradation Anaerobic microbial degradation
Nitrogen	Ammonification Nitrification Denitrification
Phosphorus	Precipitation/organic removal
Pathogens	Sedimentation Filtration Natural die-off Predation UV irradiation

Table 3.5 Monitoring results for natural wastewater treatment systems at Sveti Tomaz and Sodinci

Parameter	Sampling location	Sodinci (1100 PE)			Sveti Tomaz (500 PE)	Slovenia limit values
		October 2013	April 2014	October 2014	June 2015	
COD (mg/l)	Inflow	160	380	440	427	
	Outflow	50	60	30	<30	150
BOD_5 (mg/l)	Inflow	80	150	200	267	
	Outflow	10	13	6	<9	30

Limnos (2016)

3.10   Case Study 9: Ecoremediation Polygons, Slovenia

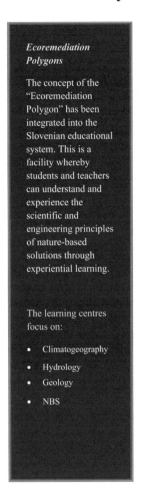

Ecoremediation Polygons

The concept of the "Ecoremediation Polygon" has been integrated into the Slovenian educational system. This is a facility whereby students and teachers can understand and experience the scientific and engineering principles of nature-based solutions through experiential learning.

The learning centres focus on:

- Climatogeography
- Hydrology
- Geology
- NBS

	Site Location:	Project Partners
 **LIMNOS Ltd** http://www.limnos.si/eng/index.php	Ecoremediation Polygon, Modraze, municipality of Poljcane, Slovenia	International Centre for Ecoremediations, University of Maribor, Ministry of Education and Sports, Limnos Ltd.

3.10.1 Site Location

This project is located in Modraze, municipality of Poljcane, Slovenia. It is an experiential learning centre (Fig. 3.34). The ecological characteristics of the polygon are varied including forest, river, natural wetlands, springs, fish ponds and constructed wetlands. The polygon is designed to get students/residents to think about environmental problems in an experiential way. The focus is to learn by doing through the hands-on field equipment and experiments available. This allows the student to understand the functions and variety of natural systems, while acquiring the skills of observation, orientation, measurement and reporting. It also promotes awareness of the role of the individual and society in the world stimulating the seeds for lifelong learning.

3.10.2 Concept

In Slovenia, a major effort is underway to integrate learning about NBS within the school and college curriculum. The International Centre for Ecoremediation from the University of Maribor together with its partners in the Ministry of Education and Sports, decided to start a project called "Establishment of the implementing conditions for experiential education for sustainable development" (Krajnc et al. 2011). They established a series of learning centres in various locations within Slovenia called "ecoremediation polygons". Ecoremediation comprises systems and processes, which function in both natural and artificial ecosystems, designed to protect and restore the environment. The polygon is a facility whereby students and teachers can "learn by doing". This educational policy has effectively created a serious of nature-based educational centres. They are designed to be interdisciplinary and innovative, where participants can learn and experience the scientific and engineering principles concerning various applications of NBS.

3.10.3 Ecoremediation Polygon, Modraze

A polygon is located in a small village Modraže in the municipality of Poljčane in the middle of a natural landscape. The function of the polygon in Modraze is to allow students to gain an experience of nature-based systems at work in both a field setting and in a classroom setting (Fig. 3.35). The demonstration site achieves this

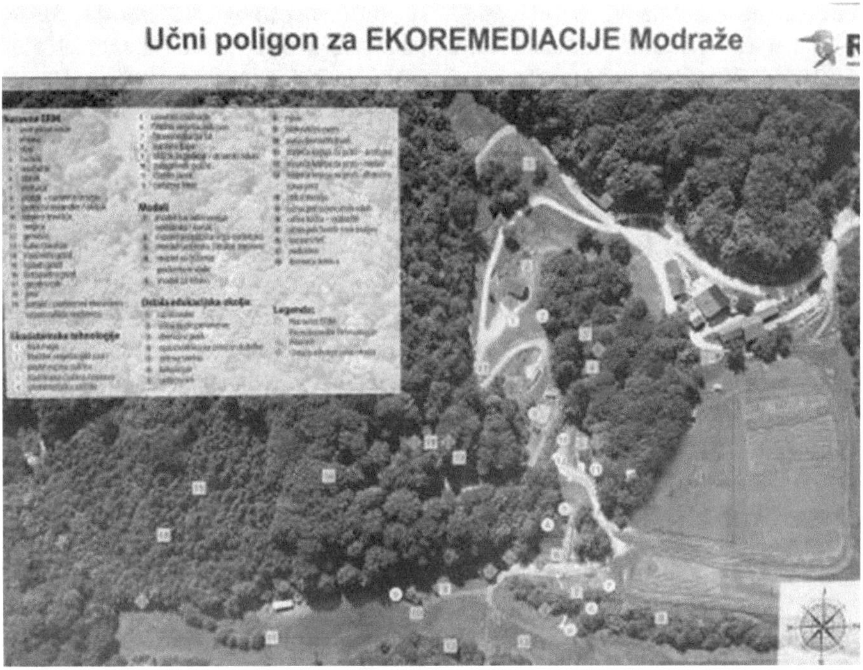

Fig. 3.34 Ecoremediation Polygon, Modraze

Fig. 3.35 Ecopod
residential units

through the use of nature as a classroom. The different ecological zones, e.g. forest, river and wetlands allow different research methods to be acquired. The constructed wetland and the innovative tactile-learning exhibits facilitate understanding of water and wastewater treatment processes, but also allow for monitoring of water quality and treatment performance. The presence of two protected species in the river system, is of special interest and stimulation.

Educational Facilities

Educational facilities within the polygon enable independent or guided learning and teaching of the laws and processes that occur in nature (ecosystems), as a basis for understanding complex physical geographical processes in the environment (Fig. 3.36). The existing forest and water ecosystem within the chosen area was retained and a residential block with cooking, toilet, classroom and laboratory facilities was constructed. A constructed wetland was added to treat the wastewater from this block. This allowed for innovative tactile demonstration models of the processes involved in water and wastewater treatment.

A nature trail has been constructed to allow students to engage with the natural environment comprising mixed deciduous forest, coniferous forest, forest edge, isolated springs, streams, river embankment and its vegetation buffer zone, dry meadow, wet meadow – wetland, farm, pond and fish pond. Tactile exhibits are set up within each ecosystem highlighting the various characteristics of each habitat. This biodiversity serves as the classroom for students (Fig. 3.37). The principles of ecology, forests, riverine flora and fauna, geography and eco-remediation are acquired by the students through mapping and sampling.

The learning environment focuses on the acquisition of knowledge and skills in a number of areas:

- **Climatogeography** – At the site temperature, wind speed, and air pressure can be measured. By observing the weather, the microclimate of a given area can be determined.
- **Geomorphology** – By collecting rocks in the area, students can train in the study of types of rocks, as well as the determination of the chemical properties of soil. Students can also observe and sketch a set of the landforms that occur on the ground.
- **Hydrology**–students can be trained in the physical and chemical analysis of water, as well as in interpreting the data. There is also a possibility to construct simple devices for monitoring the characteristics of water.
- **Phytogeography and Zoogeography**- Students can determine by themselves the dominant plant and animal species in this area. There is the facility to compare natural vegetation with anthropogenic vegetation and to identify the interactions and consequences.
- **Soil mechanics** – At the site students learn and practice the various methods of identifying soil characteristics. Through the use of various drills they learn how to take a soil sample and analyse soil properties.

Fig. 3.36 Ecosystem
training materials

Fig. 3.37 Examples of
education in biodiversity
systems

Fig. 3.38 Working model
of constructed wetland

Constructed wetlands

The constructed wetland serves to facilitate demonstration of the principles of water and wastewater treatment. Demonstration rigs are used to explain hydraulic and organic loading, and the function and structure of wetland treatment mechanisms. These demonstration rigs also allow for sampling so the techniques of sampling water and wastewater are acquired under controlled conditions. The constructed wetland experimental rigs consist of 3–4 successive chambers, which are insulated with foil and filled with medium through which sub-surfacewater flows (both horizontal and vertical) (Fig. 3.38). Students can determine, with chemical analysis, the level of water pollutants before the treatment plant and the quality of water atthe outlet of the wetland. The model design facilitates the addition of a dye to the contaminated water allowing direct observation of flow paths within the media.

3.11   Case Study 10: Domestic Rainwater Harvesting, Ireland

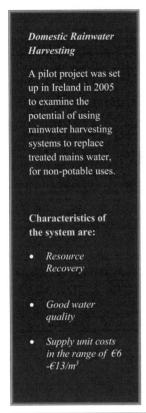

Domestic Rainwater Harvesting

A pilot project was set up in Ireland in 2005 to examine the potential of using rainwater harvesting systems to replace treated mains water, for non-potable uses.

Characteristics of the system are:

- *Resource Recovery*

- *Good water quality*

- *Supply unit costs in the range of €6 -€13/m³*

ResourceRecovery Reuse

Climate Change Adaption

	Site Location:	Project Partners
Development Technology in the Community Research Group (DTC-DIT), Dublin Institute of Technology, Ireland. www.dit.ie/dtc	Millford Park, Ballinabrannagh, Co Carlow	The project leaders were the Department of Civil & Structural Engineering, Dublin Institute of Technology (DIT), in association with the Department of Environment, Heritage &Local Government, the National Rural Water Monitoring Committee and Carlow County Council.

Fig. 3.39 Ballinabranagh housing scheme

3.11.1   Site Location

A pilot project was set up in Ireland in 2005 to examine the potential of using rainwater harvesting (RWH) systems to replace treated mains water, for non-potable uses (Fig. 3.39). The site of this project was a new housing estate 6 km south of Carlow town at Milford Park, Ballinabrannagh, Co. Carlow, at 52° 47′ 08″ N 6° 59′ 56″ W. Mains water was supplied by the Ballinabrannagh group water scheme. Initially four houses were selected for the study: one, a bungalow fitted with RWH facilities, and three 2 storey houses with standard plumbing. The three standard houses acted as controls to allow savings on mains water usage by the installed RWH system to be evaluated.

3.11.2   Concept

This study addressed concerns over harvested rainwater quality by undertaking a monitoring programme to establish the quality of harvested rainwater in an Irish context and to examine the potential of using RWH systems to replace treated mains water for non-potable uses for domestic applications. The project involved the design, installation, commissioning and monitoring of RWH facilities in a rural housing development. A monitoring program was carried out to examine the physico-chemical and microbiological quality of the harvested rainwater. A water usage monitoring program was also established to quantify the volume of water used and the efficiency of the RWH system installed. Daily/monthly rainfall, water demand, mains top-up were monitored and analysed for the domestic sites. An economic model was developed to: (i) calculate the cost of producing one m^3 of water using RWH, (ii) compare the Net Present Value (NPV) cost of RWH water supply versus mains water supply and to (iii) illustrate the preferred scenarios in Ireland under which RWH is economically viable.

3.11.3   Design Criteria

The design criteria for the RWH system was to supply 45 1 per head per day $(L^1hd^{-1}d^{-1})$ for toilet use in a four-person household with capacity for a thirty day dry storage period. This required that the system have 5.4 m^3 storage capacity.

3.11.4   Key Drivers

Water demand in Ireland is typically met by importing large volumes of water from neighbouring catchments. All mains water in Ireland is treated to drinking water quality standards. The key driver in this study was the fact that rainfall in Ireland is available throughout the year, with annual rainfall depths varying depending on whether the location is on the east of the country or the west of the country. RWH in Ireland has had limited uptake, with concerns expressed by local authorities over water quality and possible cross-contamination of the mains water system by harvested rainwater. In addition, the economics of supplying harvested rainwater compared to the cost of mains water had not been investigated. This study sets out to address these knowledge gaps.

3.11.5   Operation Characteristics

The RWH system installed collected the harvested rainwater in a 9 m^3 underground storage tank. The rainwater was pumped on demand to a header tank in the attic from which the toilets and garden tap were supplied by gravity. The RWH system collected water from roof surfaces only. Rainwater from the downpipes was diverted to an underground Rainman 1^{Tm} filter that separated solids from the rainwater. The solids were diverted to the surface water drainage system. No first flush or diversion device was installed (Fig. 3.40). All connections to the rainwater drainage system were sealed to prevent contamination from surface water. In periods of low rainfall the rainwater header tank was filled from the mains water header tank by means of a solenoid valve. A tundish type AA air gap was installed to ensure that no backflow

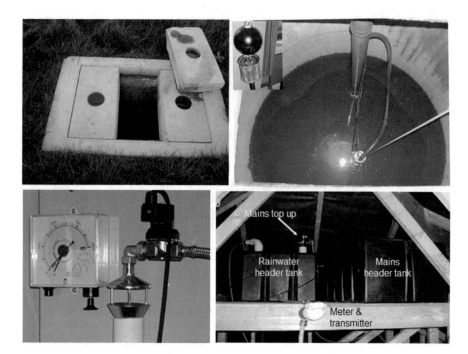

Fig. 3.40 Domestic rainwater harvesting system, Carlow

to the mains water supply could occur. A data logger system was also installed with flow monitoring to assess household water usage. A weather station was installed on site to collate weather data.

3.11.6 Natural Treatment Mechanisms

As the water passes through the various stages in the system it is exposed to processes that simultaneously reduce/eliminate the microbiological load. The rooftop provides the entry point for the majority of contaminants although parallel processes simultaneously reduce the microbiological load through UV, heat and desiccation. Within the tank, it has been shown that biofilms actively remove contaminants from the water supply. Filtration occurs as the rainwater passes through the filter with sedimentation also occurring in the storage tank. Tank water must pass through a pump and possibly through a hot water system before human contact, which impose sudden stresses on bacteria, disrupting cell structure and integrity. Each of these components influences water quality within the collection train.

3.11.7   Performance

- The rainwater harvested at the domestic installation in County Carlow was in compliance with the Bathing Water Regulations for 100% of samples taken and was of a suitable quality for use in non-potable applications.
- Results showed that the harvested rainwater complied with the more stringent Drinking Water Regulations for 37% of samples taken.
- These monitoring results represent a worst case scenario, as no first flush device was installed and no disinfection of the system took place.
- An efficient disinfection programme could have ensured that the quality of the harvested rainwater was in compliance with microbiological Drinking Water Regulations.
- A discount variable analysis on the cost of rainwater supplied versus the cost of mains water shows that harvested rainwater cost in the range of €6.81/m³ to €13.43/m³.

3.11.8   Resource Recovery Product

Roof water from this development would normally be discharged directly to a surface water network without any treatment. This is a potentially valuable resource that could be recovered to produce a product, fulfilling the basic requirements for a circular economy approach to water.

3.11.9   Consumer Acceptance

As regards installation, operation and maintenance the householder expressed satisfaction with the system and the product, i.e. water for non-potable uses. The main issue identified with regard to consumer acceptance concerned the increased cost of RWH over mains water.

3.12   Case Study 11: Agricultural Rainwater Harvesting, Ireland

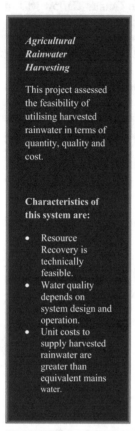

Agricultural Rainwater Harvesting

This project assessed the feasibility of utilising harvested rainwater in terms of quantity, quality and cost.

Characteristics of this system are:

- Resource Recovery is technically feasible.
- Water quality depends on system design and operation.
- Unit costs to supply harvested rainwater are greater than equivalent mains water.

ResourceRecovery Reuse

Climate Change Adaption

	Site Location:	Project Partners
 **Development Technology in the Community Research Group (DTC-DIT), Dublin Institute of Technology, Ireland.** www.dit.ie/dtc	Clonalvy, Co. Meath	The project leaders were the Department of Civil & Structural Engineering, Dublin Institute of Technology (DIT), in association with the Department of Environment, Heritage & Local Government, the National Rural Water Monitoring Committee.

3.12.1   Site Location

A pilot project was set up in Ireland in 2005 to examine the potential of using RWH systems to replace treated mains water, for non-potable uses. The agricultural site was located at Clonalvy, Co. Meath, approximately 50 km north of Dublin at 53° 34′ 54″ N 6°20′ 26″ W. Originally, a dairy farm, the farmer switched to beef production during the project. In March 2007, there were 114 cattle and 50 calves on the farm. Potable water was supplied to the farm by Meath County Council.

3.12.2   Concept

This study set out to establish the quality of harvested rainwater in an agricultural context in Ireland and to examine the potential of using RWH systems to replace treated mains water for non-potable uses for agriculture. Two distinct sampling regimes were carried out. The first, a 12-month regime, was carried out on the first installation. Modifications to this system resulted in a second sampling regime.

3.12.3   Design Criteria

The farm buildings lie in the centre of the farm and the relevant buildings to the project are two sheds/barns, each of approximately 1000 m^2 roof area (Fig. 3.41). Potable water was supplied to the farm by Meath County Council. Rainwater from the two barns was drained by gravity to an underground precast 9 m^3 concrete collection tank. The system was designed to supply non-potable water for cattle in the barn and in the troughs around the farm. The collection tank was fitted with a pump and a float switch, and the overflow pipe was connected to an adjacent field drain. The harvested rainwater was pumped via a 25 mm rising pipe to two 22 m^3 precast concrete reservoir tanks located on an adjacent elevated site. A mains top-up connection ensured mains water supply to the reservoir during periods of low rainfall. The harvested rainwater was distributed, via a 25 mm pipe, by gravity to supply the drinking troughs for cattle on the farm.

3.12.4   Operation Characteristics

Harvested rainwater was conveyed via underground pipe work to a collection tank. A 9 m^3 pre-cast concrete tank acts as an initial collection tank for the rainwater. A filtration system was needed to prevent leaves and other material being washed into the gutters thus entering the RWH system. Commercially available filtration components were sourced and installed (Fig. 3.42).

Fig. 3.41 Agricultural rainwater collection system

Fig. 3.42 Filter unit on downpipes

A Lindab leafbeater™ and a BRAE ™ filter were installed on the three downpipes conveying rainwater to the collection tank via the underground pipe work. The Lindab leafbeater filtered off any large debris from the rainwater flowing in the down pipe. As the rainwater flowed it washed any trapped debris away and out of the water stream flowing downwards towards the collection tank. The BRAE ™ filter removed any particles which passed through the leafbeater. A submersible Multigo pump was installed in the collection tank to allow pumping of rainwater up to the storage tanks. From the collection tanks the rainwater was pumped up to two interconnected 22 m³ concrete storage tanks giving a total storage capacity of 44 m³. The cattle troughs on the farm were fed from the storage tanks by gravity feed. The storage tanks are approximately 9 m above the collection tank and cattle troughs. An electronic RWH system controller was installed in the farm building adjacent to the collection tank (Fig. 3.43). This controller was connected to ball-cocks installed in the storage tanks to control the movement of water within the RWH system. Two ball-cocks were installed in the storage tanks; one controlled the infilling of rainwater from the collection tank, the second controlled the flow of mains water to top up the system. On the side panel of the control panel a red light was connected and mounted to give a quick visual check that the pump was functioning. The ballcock controlling the rainwater flow to the storage tank was set at approximately 3 m from the tank floor. It controlled the pump in the collection tank, switching it on and off as required. The second ball-cock was installed at approximately 1 m above the tank floor providing mains water back up to the storage tank. This ensured water supply to the cattle troughs during periods of dry weather when there was insufficient rainwater available or in the case that the pump failed. Gravity was used to distribute water to the farmyard troughs and some of the field troughs.

Removal Mechanisms
The rainwater was exposed to

- Thermal treatment on the barn roof,
- UV treatment on the barn roof,
- Filtration as the rainwater passed through the filter,
- Sedimentation in the collection tank,
- Sedimentation in the storage tank.

Performance
- The RWH installation in Clonalvy, Co. Meath, supplied harvested rainwater, which complied with Bathing Water Regulations.
- The physicochemical results from the site during the initial period complied with the Drinking Water Regulations over the sampling period, except for ammonia. The microbiological results breached both the Drinking and Bathing Water Regulations on all sampling dates.
- The results from the agricultural site illustrate the importance of the system design and construction, on the harvested rainwater quality. Properly engineered and constructed systems can provide a potential onsite water resource for agriculture in Ireland.

3.12.5   Resource Recovery Product

Roof water from the agriculture buildings would normally be discharged directly to a surface water network without any treatment. This is a potential valuable resource that could be recovered to produce a product, fulfilling the basic requirements for a circular economy approach to water (Fig. 3.44).

Fig. 3.43 Rainwater storage system with mains water back-up

Fig. 3.44 Rainwater consumers

3.13   Case Study 12: School Rainwater Harvesting, Ireland

School Rainwater Harvesting

The function of the project was to monitor a rainwater harvesting system for a school which minimised installation and operational costs.

Characteristics of this system are:

- Resource Recovery is technically feasible.
- On-going pumping costs are avoided.
- The system met 50% of mains water demand for non-potable uses in the school.

	Site Location:	Project Partners
 **Development Technology in the Community Research Group (DTC-DIT), Dublin Institute of Technology, Ireland.** www.dit.ie/dtc	Carrowholly National School, Westport, Co. Mayo	The project leaders were the Department of Civil & Structural Engineering, Dublin Institute of Technology (DIT), in association with the Department of Environment, Heritage &Local Government, the National Rural Water Monitoring Committee.

3.13.1   Site Location

The site was at Carrowholly National School, Westport, Co. Mayo. A RWH facility was installed in the school as part of a new extension. The school is located in the parish of Kilmeena in County Mayo at 53°48′ 52.92″, −9°.35′ 28.29″. Situated on the shores of Clew Bay and at the base of Croagh Patrick, the school is positioned three kilometres north–west of the town of Westport. It is sited on a high, level plain and the prevailing wind direction is south-westerly.

3.13.2   Concept

The function of the project was to determine the feasibility of utilising harvested rainwater to replace treated mains water, for non-potable uses in a school (McCarton and O'Hogain, 2011). The harvested rainwater was supplied to toilet cisterns.

3.13.3   Design Criteria

The RWH system was designed by Robert Kilkelly & Associates, Civil Engineering and Architectural Services, Westport, Co. Mayo. The design parameters and specifications were:

- Roof Catchment Area = 124 m^2
- RWH Tanks: Two number 682 l polyethlene coldwater cistern covered tanks, supplied by Envirocare Pollution Control, Dunmore Road, Glenamaddy, Co. Galway.
- Rainwater filters used were Koss Milk Sock Filters, supplied by Chemical Services Ltd., Chapelizod Industrial estate, Dublin 20.
- A standard ball cock was placed on the mains inlet pipe which shuts off when full and prevents syphonage. There were no non- return valves placed here.
- A meter was fitted on the inlet from the mains, which was supplied by T.C.M Controls Ltd., Greenmount Industrial Estate, Dublin 6.
- The meter on the toilets that was fed from the water harvesting tank was placed on the outlet from the water tank in the plant room.
- The cost to supply and fit the water harvesting units, including the filters, meters etc. was €7000.
- Annual cost of water to the school was a flat rate of €325.
- The 4 WCs fed from the RWH tanks were fed directly by gravity and were the only pipe work from these tanks.
- There was no pump in the RWH system.

3.13.4   Key Drivers

A key driver in sustainable water management in schools is maximising existing resources to augment mains water supply for non-potable uses. The school in this pilot study had green flag status, which signified the school's efforts to promote environmental awareness. The RWH facility was a component of the school's activities in promoting awareness and implementing sustainable water management practices.

3.13.5   Operation Characteristics

The roof catchment area draining to each system was 62 m², providing a total catchment area of 124 m². The roof covering was a Sarnafil® membrane. The roof was pitched at approximately 10° resulting in rainwater draining to the roof edge, which had a 150 mm deep perimeter. The roof perimeter gutters channelled the rainwater directly to the storage tanks located on each side of the roof space. Each tank had a storage capacity of 682 l and was fitted with a lid. A first flush device was not fitted to either RWH system. Rainwater was directed straight through to the storage tank via a 50 mm pipe. A 100 mm diameter overflow pipe drained to the external roof channel. A cloth geotextile filter was fitted on the inlet with 4 mm aperture to capture any fines from the roof surface. Mains water was piped to the rainwater storage tank. A ball-cock valve controlled the mains water inflow. The level of the ball-cock was set below the level of the intake from the rainwater. During periods of low rainfall intensity, when the level of water stored in the tank fell below this critical level the mains water supply valve opened. Conversely, during period of high rainfall intensity, the mains supply valve closed and water was supplied by rainwater only. This top up system ensured reliability of supply. The rainwater storage tank had a secondary overflow system comprising two 50 mm pipes. There were no pumps in this system. Rainwater from the roof drained by gravity to the storage tanks. Supply from the storage tanks to the building was also by gravity (Fig. 3.45).

3.13.6   Performance

Results showed that 56% of non-potable water demand was met, thus the installation could be said to be 56% efficient.

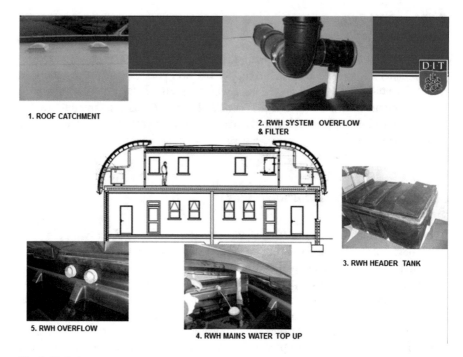

Fig. 3.45 Rainwater system for school

3.13.7   Resource Recovery Product

Roof water from the school buildings would normally be discharged directly to a surface water network without any treatment. This is a potential valuable resource that could be recovered to produce a product, fulfilling the basic requirements for a circular economy approach to water. In this installation, harvested rainwater was collected, filtered and used to augment mains water supply to toilets within the school.

3.13.8   Community Involvement

The green school initiative is an international environmental education programme, environmental management system and award scheme that promotes and acknowledges long-term, school action for the environment. Students and teachers participate in a range of actions from waste management to energy to water. The school was involved in the operation and monitoring of water usage through this scheme.

3.14   Case Study 13: Zero Discharge Natural Wastewater Treatment, Ireland

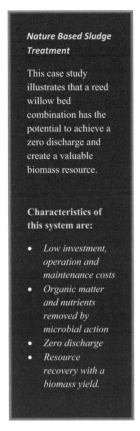

Nature Based Sludge Treatment

This case study illustrates that a reed willow bed combination has the potential to achieve a zero discharge and create a valuable biomass resource.

Characteristics of this system are:

- *Low investment, operation and maintenance costs*
- *Organic matter and nutrients removed by microbial action*
- *Zero discharge*
- *Resource recovery with a biomass yield.*

ResourceRecovery Reuse

Water / Wastewater Treatment

Restoring Ecosystems

Development Technology in the Community Research Group (DTC-DIT), Dublin Institute of Technology, Ireland. www.dit.ie/dtc

Site Location:	Project Partners
Lynches Lane, in the administrative area of South Dublin County Council (SDCC).	The project partners were South Dublin County Council (SDCC) and Dublin Institute of Technology (DIT).

3.14.1 Site Location

This project was a 2-year study to monitor the performance of a hybrid reed bed/ willow bed facility at Lynches Lane, in the administrative area of South Dublin County Council (SDCC).

3.14.2 Concept

Reed Bed Wastewater Natural Wastewater Treatment
Reed beds have been used for the last 50 years to treat wastewater, in Europe (Vymazal 2005). The design has evolved from horizontal-flow reed beds through vertical beds to hybrid beds, and latterly compact vertical flow beds (Weedon 2003). The principal mode of treatment is a combination of sedimentation, filtration, aerobic/anaerobic degradation, ammonification, nitrification/denitrification, plant uptake and matrix adsorption (Brix 2004). The efficiency of vertical-flow beds over horizontal-flow beds in treating wastewater has seen their use increase, especially over the last 10 years.

Willow Bed Wastewater Treatment
Denmark was one of the first countries in Europe to conduct research into willow wastewater treatment systems. The Danish research prompted pilot studies in other countries. The purification efficiency of willow treatment systems has been demonstrated in several countries (Borjesson 2008; Perttu 1999). Performance of systems has varied depending on site specific conditions, influent, design and operational and maintenance regimes. Some general performance characteristics of properly designed willow treatment systems can be defined as follows (Brix 2006)

- The system can be designed to facilitate all wastewater being evaporated to the atmosphere on an annual basis.
- Nutrients and heavy metals are removed by harvesting the willows (or accumulate in the bed).
- Sizing of beds is determined by the difference between precipitation and evapotranspiration.

3.14.3 Key Drivers

The site is a kilometre away from the nearest mains sewer. The mission of the SDCC is to achieve environmental excellence in all its projects. The option of storing wastewater on site and tankering it to a wastewater treatment facility would have incurred on-going operational costs. The adoption of a NBS also avoided on-going operational costs and the requirements for chemicals to be stored on site.

3.14.4   Facility Design and Installation

This hybrid reed and willow bed sewage treatment system (HWTS) currently services the Parks department depot at Grange, Lucan, Co. Dublin. The initial system was commissioned in 2002. It was designed for a PE of 15. This resulted in a design flow of 3.0 m^3 day^{-1}. Three sites are served by the system, a local authority depot, a private house and a travellers' halting site. A willow bed tertiary filter system was installed in 2008. The wastewater flows by gravity to a septic tank. From here it overflows to a pump sump, where it is pumped to the HWTS. The wastewater flows by gravity through the system (Fig. 3.46).

The vertical beds were sized at 2 m^2 PE^{-1}, to achieve BOD removal and complete nitrification on two vertical stages (Cooper et al. 1996). The beds were lined with a high-density polyethylene liner. An overall depth of media of 0.6 m comprised two bottom layers of 15 cm each, 20 cm of 6 mm diameter washed pea-gravel and 10 cm sharp sand layer. The sand was selected using the Grant method, with a test value of 45 s (Cooper et al. 1996).

3.14.5   Monitoring Regime

The reed bed was monitored for 2 years. Samples were taken aseptically at four points within the system. The physico-chemical analysis tested for nitrate, ammonia, Kjeldahl nitrogen, pH, total suspended solids (TSS), orthophosphate, chemical oxygen demand (COD) and biochemical oxygen demand (BOD). Samples for microbiological analysis were taken in sterile bottles to ensure no cross- contamination. They were analysed for the time dependent parameters, Coliforms and *E. Coli*. All analysis of water quality parameters was carried out in an Irish National Accreditation Body (INAB) accredited laboratory as per Standard Methods. Figure 3.48 shows the system during winter 2012. Figure 3.47 illustrates the tipping bucket system which avoids the need for a pump (Fig. 3.48).

3.14.6   Performance

A fortnightly programme was set in place over the 2 years of the project. This monitored clogging of reed beds, odour, flow and pipework including tipping buckets, pumps, reeds, willows, pipework. Table 3.6 shows the overall performance characteristics for the system.

Fig. 3.46 Overview of treatment system showing PVF reed bed, SVF reed bed, HF reed bed and Willow bed

Fig. 3.47 Tipping bucket distribution system

Fig. 3.48 System operating in winter 2012

3.14.7   Reed Bed Performance

Removal values for COD and BOD were comparable with results achieved at other hybrid systems built in Ireland. Suspended solids removal was slightly lower, at 85%. Coliform and E.coli removal rates were also marked at 94% respectively.

Willow Bed Performance
No outflow was observed from the willow bed during the monitoring period. There were frequent periods when the willow bed was dry throughout. This left three possible pathways for the effluent. These were passage through the soil, absorption to the roots and evapotranspiration of the wastewater, and or evaporation in the open trenches due to climatic factors such as wind and sunlight. To determine percolation through the soil, a series of tests were performed. The average permeability of the samples was 2.3×10^{-7} m/s. From this, we may conclude that the wastewater is being removed primarily by evapo-transpiration effects. During the 2-year monitoring period, the Reed bed/willow bed system at Lynches Lane Co. Dublin achieved zero discharge (O'Hogain and McCarton, 2011).

3.14.8   Resource Recovery Product

The system produces a resource in the form of willow biomass. A total of 180 willow cuttings were planted in February 2008. Three willow varieties were planted namely *Salix triandra*, *S. purpurea* and *S. Viminalis*. In winter 2009, a biomass audit was carried out. The biomass audit determined the average plant height to be 1.9 m with a range of 1–3.1 m. Stem thickness ranged from 6 to

Table 3.6 Overall treatment performance

Parameter	Influent (Sewage)					Horizontal flow reed bed (Effluent)					Percentage removal		
	Mean	S.D.	Median	Min	Max	Mean	S.D.	Median	Min	Max	Mean	S.D.	Median
NH_4_N (mg l^{-1})	45	40	28	3	144	13	12	12	0	41	71	71	56
KJN as N (mg l^{-1})	44	31	35	4	117	13	12	11	0	45	70	71	69
NO_3_N (mg l^{-1})	5	10	1	0	40	34	47	4	0	19	−586	−353	−418
PO_4P (mg l^{-1})	4	4	2	0	14	2	1	2	1	6	45	73	17
Coliforms (MPN/100 ml)	2,389,261	4,808,118	1,198,000	1	24,196,000	133,685	402,744	10,462	31	1,986,300	94	92	99
E.coli (MPN/100 ml)	711,074	1,459,254	233,300	1	7,270,000	40,572	121,023	1710	10	579,400	94	92	99
COD (mg l^{-1})	289	206	327	33	890	37	26	32	0	105	87	88	90
BOD_5 (mg l^{-1})	129	99	136	4	316	11	15	4	1	58	91	85	97
TSS (mg l^{-1})	101	117	60	17	524	15	22	9	1	96	85	81	85
pH	8	1	8	7	12	7	0	7	7	8	–	–	–

24 mm with an average thickness of 14 mm. The resulting biomass could be harvested and then utilised for heat and electricity production. 2500 ha of miscanthus (miscanthus giganteus). Further studies are required in Ireland on the application of willow systems in different climatic regions and on the capability of achieving zero discharge (EPA 2016).

3.15   Overall Performance Characteristics of NBS Case Studies

The various case studies illustrate the application of NBS across a variety of water sources. The water sources include rain water, surface water, ground water, wastewater, sludge, flood water, industrial effluent, drainage water, agricultural

Table 3.7   Principal contaminants and removal mechanisms in NBS case studies

Contaminant	Removal Mechanisms	1. Inland Shore	2. Managed Aquifer Recharge	3. Green Port	4. Can Cabanyes	5. STW	6. Mill? Ponds	7. Nutrient Recovery	8. NBS Slovenia	9. Ecoremediation, Slovenia	10. RWH - Domestic	11. RWH - Agriculture	12. RWH - School	13. Zero Discharge WWT
Gross Solids	Screening	■	■		■	■	■		■	■	■	■	■	■
	Filtration	■	■		■	■	■		■	■	■	■	■	■
Suspended Solids	Sedimentation	■	■	■	■	■	■		■	■		■	■	■
	Settlement	■	■	■	■	■	■		■	■		■	■	■
Dissolved Organics	Aerobic /	■			■	■		■	■	■				■
	Anaerobic Digestion.	■			■	■		■	■	■				■
Nitrogen	Ammonification	■	■		■	■		■	■	■				■
	Nitrification	■	■		■	■		■	■	■				■
	Denitrification													
Phosphorous	Precipitation							■						
	Organic Removal	■	■		■	■		■		■				■
	Recovery	■												■
Pathogens	Sedimentation	■	■	■	■	■			■	■	■	■	■	■
	Filtration	■	■	■	■	■			■	■	■	■	■	■
	Natural die-off	■			■	■			■	■	■	■	■	■
	Predation	■			■	■			■	■				■
	UV Irradiation	■			■	■			■	■				■

runoff and lake water. The predominant contaminants and the removal mechanisms required to improve water quality, are given in Table 3.7. NBS incorporate these removal mechanisms either naturally or by design into their structure. The physical removal mechanisms, such as settlement/sedimentation and filtering tend to occur in all NBS, in particular where influent velocity is reduced. Biological treatment also occurs in most NBS systems, though the majority would tend to feature aerobic treatment, as they are amenities and anaerobic conditions would only occur locally within the system as in Case Studies 1 and 4. Nutrient removal can also occur naturally or can be part of the NBS system as in Case Study 7. Pathogen removal is a function of residence time and structure of the NBS which tend to be in the open air and exposed to sunlight. Residence time and exposure to sunlight can prove effective in removing pathogens, but the low trophic status of most NBS mean that there is organic material for pathogen survival (Table 3.7).

Chapter 4
Reclaimed Water

Abstract This chapter presents a definition of reclaimed and reused water. It also presents an overview of potential uses for reclaimed water in terms of agricultural re-use, environmental re-use, groundwater recharge, non-potable water re-use. This section will also consider the circular economy of water (CEW) and define the term "fit for purpose". Finally it will present an overview of European and Global Water Reuse guideline and applications.

Keywords Reclaimed Water · Reused Water · Water Reuse Guidelines

4.1 What Is Reclaimed Water

4.1.1 The Circular Economy of Water (CEW)

Water infrastructure has traditionally adopted a linear model, where water is abstracted at source, treated, used and treated again prior to disposal to the environment through either surface water or groundwater regimes. In the linear approach to water, products are disposed of after use. The circular economy is by design restorative of ecosystems. The circular economy, operating within planetary boundaries, is waste-free and resilient. The characteristics of the circular economy of water involve an approach where water and its contents, along the entire supply chain are viewed as a resource to be recovered and reused. NBS are an inherent component of this model. This approach will result in a multi-functional solution which is resilient, particularly to the challenges faced in a post climate change environment and that is restorative of ecosystems as illustrated in Fig. 4.1.

4.1.2 Reclaimed Water

Typically in the linear economy raw water is abstracted and treated to potable (drinking) water standards. This is then supplied to users, both domestic and industrial who use it for various purposes. Some of these uses require advanced treatment

© Springer International Publishing AG, part of Springer Nature 2018   95
S. O'Hogain, L. McCarton, *A Technology Portfolio of Nature Based Solutions*,
https://doi.org/10.1007/978-3-319-73281-7_4

Fig. 4.1 The circular
economy of water

to produce water of a higher quality. The use of the water changes (reduces the quality) and therefore it has to be treated again before it can be discharged back into the environment. Water after use is typically termed "wastewater" but more accurately should be termed "used water". ***Reclaimed water is used water that has undergone treatment to upgrade the quality to enable it to be reused.***

4.1.3   Fit for Purpose Model

A portfolio of engineered and natural treatment options exist to mitigate microbial and chemical contaminants in reclaimed water facilitating a multitude of reuse options (Multiple Waters). The treatment level required depends on the end use and is focused on protecting public and environmental health. This concept can be considered as "Fit for Purpose" where the level of treatment produces a quality of water equal to or above that required for the intended reuse function of the water. This fit for purpose model will result in multiple waters for multipe uses where the function of the water governs the water quality. This is illustrated in Fig. 4.2.

If we apply a fit for purpose model to the water management system it is possible to create a system where used water is reclaimed to different quality levels suitable for reuse within various applications.

4.2   Potential Uses for Reclaimed Water

There are two major types of water reuse: direct reuse and indirect reuse.

- **Direct Reuse** is typically considered as the introduction of reclaimed water directly from a water reclamation facility to a water distribution system.
- **Indirect Reuse** can be considered as the placing of reclaimed water into storage (lake, river or aquifer) where it can be abstracted to be used again through conventional water treatment and distribution systems.

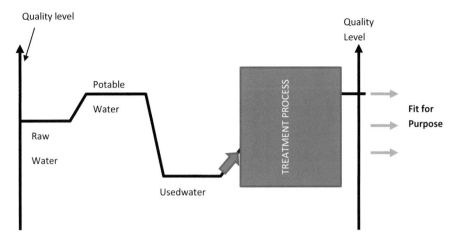

Fig. 4.2 Multiple waters for multiple uses can achieve desired levels of water quality to suit reuse (Adapted from USEPA 2012)

Typically divide water into two principle quality levels, potable water (suitable for drinking) and non-potable (suitable for other uses). This facilitates a subdivision of possible reuse of reclaimed water as follows;

Class 1: Direct Potable Reuse – This is reclaimed water that is treated and discharged from a wastewater treatment facility directly to the potable water network.

Class 2: Indirect Potable Reuse – This is reclaimed water that is discharged directly into water reservoirs (storage) and aquifers that are then used as raw water abstraction points to drinking water treatment facilities which treat the water to potable standards.

Class 3: Direct Non-Potable Reuse – This is reclaimed water that is used directly after treatment for non-potable uses either on site or off site.

Class 4: Indirect Non-Potable Reuse – This is reclaimed water that is discharged from the treatment facility to a water body (storage) that is then used as a water source for non-potable uses.

Use of reclaimed water may have other benefits, such as recycling of nutrients and energy savings. The contexts for the use of reclaimed water vary significantly across the world. Therefore, there is no 'one size fits all' approach which would be appropriate.

The NBS technology portfolio within this book has outlined some applications of reuse of water in the following areas:

- **Urban Reuse**
- **Agricultural Reuse**
- **Environmental Reuse**
- **Ground Water recharge**
- **Non-Potable Reuse**

Table 4.1 Water reuse matrix

Water reuse applications	NBS case studies												
	1. Inland shore	2. Managed aquifer recharge	3. Green port	4. Can Cabanyes	5. STW	6. Mill ponds	7. Nutrient recovery	8. NBS Slovenia	9. Ecoremediation, Slovenia	10. RWH - domestic	11. RWH - agriculture	12. RWH - School	13. Zero Discharge WWT
Urban reuse. Recreation Irrigation													
Agricultural reuse Food chain Non food chain													
Environmental reuse Wetlands Ecosystem services Hydraulic infrastructure													
Ground water recharge													
Non-potable reuse Harvested rainwater Grey water													

Table 4.1 shows the water reuse matrix for the NBS portfolio.

4.2.1   Agricultural Reuse

Agriculture is the main water use in many countries and in some cases it accounts for up to 80% of all freshwater abstractions, i.e., Spain. Agricultural water reuse is generally regarded as irrigation water and can be considered for use in the food chain and with non-food chain products. Higher water quality standards are required if the water reuse is for products entering the food chain. The mechanisms used to distribute the water are also regulated if food production is involved. Reclaimed water can also contain levels of nitrogen and phosphorus which could potentially reduce the need for additional application of mineral fertilisers.

4.2.2 Environmental Reuse

Environmental reuse covers a variety of applications, designs and functions. Wetlands can be used for wastewater treatment, water treatment, storm water treatment as well as wild life habitat, fisheries habitat and biodiversity promotion. They also play a role in ecosystem services as the benefits of wetlands are benefits to humans also. Wetlands can serve many hydraulic infrastructure uses also, from storm water collection, to water storage to the inland shore discussed in Case study 1.

4.2.3 Groundwater Recharge

Groundwater recharge has been used to restore underground aquifers that have been over used. Case study 2 in the NBS technology portfolio shows an example of aquifer recharge from Barcelona for rainwater and for reclaimed wastewater with the use of a unique mix of local clay and compost. Both surface waters and groundwater are governed by legislation (i.e. The European Communities (Good Agricultural Practice for Protection of Waters) Regulations 2010).

4.2.4 Non-potable Reuse

Non-potable reuse refers to all water use except that for drinking and life support. It is normally divided into harvested rainwater and grey water. Harvested rainwater can potentially supply toilet use, washings machines and outdoor use. Grey water is referred to as used water from sinks, baths, showers and washing machines. With treatment this water can be used as carriage water domestically and grey water recycling is a common domestic reuse application. Further guidance on the design, installation and operation and monitoring of RWH and greywater reclamation systems can be sourced in the British Standards: BS8515:2009 RWH Systems – Code of Practice (BS8515 2009) and BS 8525-1:2010 Greywater systems – Code of Practice (BS8525 2010). Industrial reuse would also fall under this category. Industrial water use is determined by the quality requirements of the industry or process.

4.3 International Water Reuse Applications

4.3.1 The Aquarec Project

The AQUAREC project on "Integrated Concepts for Reuse of Upgraded Wastewater" was funded by the European Commission within the 5th Framework Programme. The project aimed to define criteria to assess the appropriateness of wastewater reuse concepts in particular cases and to identify the potential role of wastewater reuse in the context of European water resources management. In their policy brief they concluded that if utilisation of reclaimed wastewater is not to contradict the "whenever appropriate" guidance of the Urban Wastewater Treatment Directive, a definition of 'appropriateness' is needed. In other cases, they concluded that switching from conventional water resources to reclaimed wastewater is primarily hindered by cost arguments. This would demand the establishment of water prices that reflect the full-cost recovery principle on the one hand, and the monetarisation of the potential environmental benefits of wastewater reuse, on the other.

4.3.2 US Agricultural Reuse Standards

The use of reclaimed water from agriculture has been widely supported by regulatory and institutional policies in the United States. In 2009, California adopted both the "Recycled Water Policy" and "Water Recycling Criteria." Both policies promote the use of recycled water in agriculture (SWRCB 2009; California Department of Public Health 2009). The California Water Recycling Criteria require stringent water quality standards with respect to microbial inactivation (total coliform <2.2 cfu/100 mL). In 2012 the USEPA issued updated guidelines for water reuse providing comprehensive information on different water reuse practices including international aspects (USEPA 2012). Globally, the US EPA Guidelines for Water Reuse has had far-reaching influence with many countries either referencing the document or adopting the guiding principles.

4.3.3 WHO Guidelines for the Safe Use of Wastewater, Excreta and Greywater

The WHO Guidelines for the Safe Use of Wastewater, Excreta and Greywater: Volumes 1 – 4, widely adopted in Europe and other regions, is a science-based standard that has been successfully applied to irrigation reuse applications throughout the world (WHO 2006). The World Health Organisation (WHO 2006) 3rd edition

Table 4.2 National water reuse criteria

Member state	Type of criteria	Comment
Belgium: Flemish Regional Authority	Aquafin proposal to the government (2003)	Based on Australian EPA guidelines
Cyprus	Provisional standards (1997)	Quality criteria for irrigation stricter than WHO standards but less stringent than California Water Recycling Criteria (Total Coliforms <50/100 ml in 80% of cases)
France	Arrete du 2/10/2010	Refers to water reuse for agricultural and urban landscapes. Allows the use of sprinkler irrigation with minimum distances to sensitive areas. Associates water quality levels to crop to be grown. Specify distances between irrigated plots and public areas.
Portugal	Portugese National Standard NP4434:2005	Chemical and microbiological treated wastewater parameters and treatment levels are defined to protect public health, environment and crop protection.
Greece	Ministerial Decision 145116/2001	The guidelines are based on three kinds of water use: restricted, unrestricted and urban uses. For each use, a different minimum treatment is required and a different set of limit values for water characteristics is established.
Italy	Decree of Environment Ministry 185/2003	Quality requirements are defined for the three water reuse categories: agriculture, non-potable urban uses and industrial uses. Possibility for regional authorities to change some parameters or implement stricter regional norms.
Spain	Royal Decree 1620/2007	Quality Criteria defined for five sectors: residential, agricultural, environmental, industrial and recreational.
Regional Authorities of Andalucia, Balearic Islands and Catalonia	Guidelines from Regional Health Authorities	Developed their own guidance documents based on the WHO 2006 guidelines concerning wastewater recycling for irrigation use.

WssTP (2009)

of the guidelines are based on a risk assessment. These set the standards and reduction goals for pathogens and chemical parameters and give recommendations for reduction measures to achieve human health and environmental protection. The guidelines specify treatment processes, water quality standards, and monitoring regimes that minimise risks for use of reclaimed water for irrigation of crops that are ingested by humans.

Several other European countries have adopted specific water reuse criteria as summarised in Table 4.2.

The early adopters in urban water reuse projects tend to be in the USA, Japan, Australia and China. There are a number of NBS in this sector where water is reclaimed and reused in the urban context for parks and street cleaning. The role of

NBS to deliver planned indirect potable reuse (after blending with other sources) is summarised by the WssTP report entitled Sustainable Water Management inside and around large Urban areas (WssTP 2009). These include:

- Infiltration basins: ponds which allow water to infiltrate into an underlying aquifer
- River bank filtration; extraction of groundwater from a well near a river to induce infiltration from the river to the well thereby improving the quality of recovered water
- Dune filtration: infiltration of water from ponds constructed in dunes
- Aquifer storage and recovery: the storage of water in a suitable aquifer and the recovery of water when needed.

One of the longest running aquifer storage and recovery schemes is in Orange County, California where reclaimed municipal wastewater is treated by microfiltration combined with reverse osmosis and UV and hydrogen peroxide disinfection and used to replace 30% of the water withdrawn from the aquifer. The "Groundwater Replenishment System" which began in 2008 provides enough recycled wastewater to meet the needs of 850,000 orange county residents. The resulting program has been publicised as "*toilet to tap*"(Deshmukh and Steinbergs 2006).

Chapter 5
Constraints and Barriers to the Adoption of NBS

Abstract This chapter examines some of the key bottlenecks and barriers related to NBS for water resources management. Considerations here will include public acceptance and understanding of water reuse, contaminant risk, demonstration projects, financing water reuse projects and legislation/policy together with water re-use management.

Keywords Water reuse public acceptance · contaminant risk · financing

5.1 NBS Water Reuse Constraints

A detailed assessment of the NBS case studies presented in Chap. 3 illustrates some general trends regarding constraints and barriers to the adoption of NBS.

NBS, by their nature, adopt a multidisciplinary approach to water resource management (Brink et al. 2012). The basic tenet of the NBS approach is to achieve added value through ecosystem services, i.e. the benefits that accrue to humans from nature. The NBS approach gives equal importance to the scientific/engineering performance of both the infrastructure and the ecosystem services provided. The NBS approach requires a change in mind-set on the part of the professions (Architects, Engineers, Planners, etc.) and State Agencies to incorporate this multidisciplinary approach into the initial project concept.

NBS also include the local community in defining the problem and exploring feasible solutions (De Vriend and Van Koningsveld 2012). The community are considered central to both the design team and operational and maintenance team. Their local knowledge and experience make them a source of invaluable information. The existing top down approach to infrastructure provision often only considers community input when the preliminary design has been completed and provides minimal opportunity for change.

NBS are site specific, in that the solution/s are unique and incorporate local characteristics and design features that do not necessarily replicate (Deltares 2015b). This presents a challenge to the traditional approach to engineering design which seeks to standardise solutions according to strict codes of practice. The international experience presented within the case studies in this report suggests that one should seek to replicate the NBS methodology rather than the site specific solution.

Table 5.1 Constraints and Barriers to the adoption of NBS

Water reuse constraints in agriculture	Water reuse constraints in urban, environmental, groundwater and rainwater harvesting
Public acceptance and understanding	**Public acceptance and understanding**
Use of reclaimed water is perceived to change crop production rates	The success of developing countries in providing a dependable, inexhaustible supply of safe water at low cost to households has effected the willingness of the public to contemplate alternative reuse schemes
Public awareness for health risks	Public acceptance dictates that cost of non-potable water should be less than that for potable water
Retailer perception	**Contaminant risk**
Legislation/policy and management	Perceived risk of contaminants requiring additional montoring programs which are time consuming and expensive
Lack of incentives to us reclaimed water while sufficiency in conventional water resources	**Financing reuse projects**
Lack of common water reuse regulations for agriculture in Europe	Supply of non-potable water including the costs of collection, treatment, storage and distribution are often comparable or greater than for potable water schemes.
Water Framework Directive sets the principles to achieve sustainable water governance but not the means	Dual distribution networks increase costs
	Low profit margins make it commercially unattractive for utility companies
	Need to develop new business models
	Need to Cost added value (for. Ex ecosystem services)
Demonstration	**Legislation/policy and management**
Feasibility of large scale direct use of nutrient rich water for crop irrigation is not yet demonstrated	Water governance and legal frameworks immature in Europe to facilitate and regulate urban reuse projects
	Commercial actors nervous to participate in an area with poor legal clarity
	Commercial actors nervous to "learn by doing" which limits creativity and innovation within the sector
	Lack of skills and competencies along the entire supply chain from design, planning, regulators etc.

Another constraint to the adoption of NBS solutions is legislation, and more specifically the way legislation is currently framed. Used water is considered a waste. Water legislation puts the onus on "how not to" and discharge limits are based on environmental and/or public health considerations. This study has demonstrated that NBS can be used to remove contaminants and return water to a quality which can then be considered "Fit for Purpose".

Financial models also require readjustment to fully evaluate the costs and benefits of adopting NBS. The proper costing of added value, whether it is an amenity or an aesthetic value of ecosystem services, or the benefits in terms of climate resilience requires further research. Another financial barrier involves the return on investments for companies, especially SMEs, investing in NBS projects. The international experience suggests that profit margins tend to be low, thus rendering NBS commercially unattractive.

The lack of sufficiently large scale NBS demonstration projects with monitoring is a significant constraint. Internationally, pilot demonstration sites serve to promote innovation within the NBS approach and provide data on local and regional performance to assist in replication of the methodology.

Table 5.1 further summarises these constraints according to the reuse potential.

Chapter 6
Hybrid Infrastructure: Local, Regional and Global Potential of Nature Based Solutions

Abstract The traditional approach to water supply and wastewater treatment involves centralized systems designed almost exclusively at protecting community health. Future systems will focus on the provision of an integrated service to customers. Advancing innovations in water efficient strategies will be an integral component of this new water sector. This chapter presents case studies at local, regional and global level which illustrate a new approach to water management. The case studies presented include the Rain Cities of South Korea, Philadelphia Clean City Clean Waters program and Singapore Four Taps National Strategy. NBS will form an integral component of future water infrastructure that comprises a mix of high tech human built engineered (Grey) infrastructure and NBS (Green) infrastructure. This combination of approaches can be termed "Hybrid Infrastructure". These case studies illustrate the application of this hybrid infrastructure system. This hybrid infrastructure system combines centralised and decentralised systems to optimise the reclamation of water for reuse in a fit for purpose model.

Keywords Hybrid Infrastructure · Rain Cities of South Korea · Philadelphia Clean City Clean Waters program Singapore Four Taps

6.1 Traditional Engineered Water Infrastructure System

Currently, local water demand is typically met by importing large volumes of potable water from centralised water treatment facilities to decentralised water users. The water supplied is single quality (potable) without regard to the required water quality at the point of use. Simultaneously, rainwater is typically discharged unused via expensive storm water drainage systems. Current wastewater treatment systems involve decentralised collection, centralised treatment facilities and discharge to either ground or surface water systems. This traditional system is unidirectional, running from source to the site of use to disposal, without any loops or recirculation and/or recycling. The resilience and sustainability of these traditional systems, particularly given the implications of climate change, is uncertain (Fig. 6.1).

© Springer International Publishing AG, part of Springer Nature 2018   105
S. O'Hogain, L. McCarton, *A Technology Portfolio of Nature Based Solutions*,
https://doi.org/10.1007/978-3-319-73281-7_6

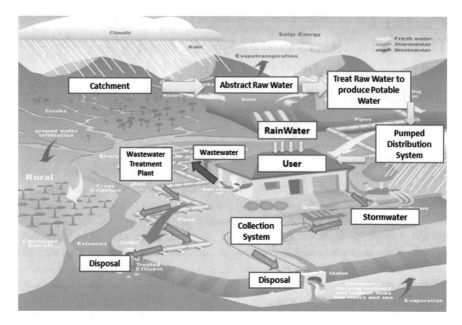

Fig. 6.1 Traditional engineered water infrastructure system

6.2   The Circular Economy of Water (Recover, Reuse, Recycle)

This system seeks to maximise recovery of valuable products from within the water stream in addition to maximising reuse and recycling options at all levels within the system. In this system rainwater is harvested, treated and reused on site. Treated rainwater is recycled back into the house for both potable and not potable use. In this system we do not use the term wastewater, rather the term "used water". Used water is collected and carried to a resource recovery facility. Here the used water is reclaimed and the improved quality allows for its reuse earlier in the process, i.e. either in the potable water treatment process, in the domestic water use, for agricultural use or for industrial use or ground water recharge. Excess water is discharged to the local water source or to groundwater. Valuable nutrients and other resources can also be mined from the used water stream for reuse. This system produces multiple waters for multiple uses (Fig. 6.2).

6.3   Hybrid Grey and Green Infrastructure

NBS will form an integral component of a future water infrastructure that comprises a mix of high tech human built engineered (Grey) infrastructure and NBS (Green) water infrastructure. This combination of approaches can be termed "Hybrid Infrastructure". This approach will require us to rethink and redesign the current

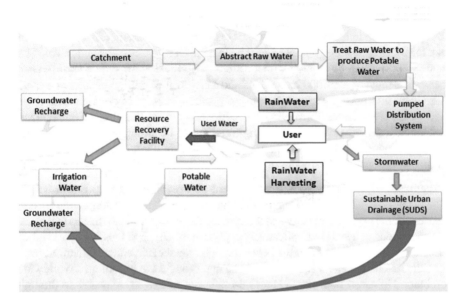

Fig. 6.2 Circular Economy of Water (Multiple waters for multiple uses)

system both by combining centralised and decentralised water treatments and by optimising the exploitation of alternative water sources in a circular economy approach to water. New localised and regional closed loop systems will ensure that used water from point of use systems is collected and carried to a resource recovery facility. Here the used water can be reclaimed and the improved quality allows for its reuse. Hybrid centralised and decentralised systems will enable a "Fit for purpose" concept to be applied to multiple waters, using different water qualities for multiple uses, depending on the local availability and user needs. This will facilitate the development of a more holistic water portfolio that utilises the benefits of both engineered and natural ecosystems. In this future system NBS can contribute to developing a robust, flexible and resilient water infrastructure at a local, regional and global level.

Unlocking the environmental, social and economic benefits of a hybrid infrastructure by combining a range of measures to store, reuse and slowly dissipate rainwater, reducing flooding and helping to conserve precious water resources has to date been viewed as aspirational. However, three examples, Star City, South Korea, Singapore and Philadelphia illustrate approaches that show that sustainable water infrastructure is not just about water management. When viewed in a holistic manner, hybrid infrastructure can involve the creation of sustainable community spaces, urban regeneration, physical infrastructure, job creation and biodiversity.

6.4   Hybrid Infrastructure: Case Study at Local Level

The potential to integrate NBS into water resources management can be considered at a local, a regional and global level. The local level refers to cities and towns. The same solutions can be applied with a difference in scale being the main difference between a town of say 4000 inhabitants and a city of four million people. The same principles of working with a spectrum of professionals, planners architects engineers etc., listening and involving the community from the start of the project, designing water in rather that designing water out and building nature in rather than building in nature all apply. A technology portfolio similar to that presented earlier in this publication is also required.

Decentralized water resource management and reuse – Star City South Korea
The Local Level: The massive migration to the cities, which characterised the closing decades of the twentieth century and the opening decades of the 21st, means that water resource management is most effectively applied to the city. In discussing hybrid infrastructure at a city level, one of the better examples are the rain cities of South Korea.

Climate change has meant that flooding and drought are becoming a worldwide problem. However, both flooding and drought are related to rainwater so through water resource management both can be mitigated. This can be achieved through the collection of rainwater before it impacts on the surface water/sewerage collection system, i.e. the drainage system. The authorities in South Korea adopted a plan to create Rain cities, based on scientific and engineering data and principles. This was aimed at Climate Change adaption through the promotion of Rain Cities (Han 2011).

Rainwater is high quality water and most of the contamination of rainwater can be said to occur after it has fallen and come into contact with the ground and travelled to a drainage system. Therefore impoundment and use on site reduces the need for treatment and transport and impacts on climate change through reduced carbon emissions and reduced infrastructure needs. The drainage collection system, be it combined or separate, is designed for a certain volume of rainfall. However climate change has resulted in more intense rainfall events where the capacity of the drainage system is unable to cope with the volume of water. Rainwater tanks and storage devices can be designed to increase the capacity of existing sewer systems without reconstruction of the sewer system.

Faced with the consequences of climate change, increased flooding and water supply problems, the South Korean Ministry of Land, Transport and Maritime Affairs (MLTM) MTML announced that the rain cities were to be designed such that rainwater is collected rather than running to surface/sewerage drains. This policy of a multipurpose system has the dual purpose of flood mitigation and water conservation. It is also proactive utilising a remote control system which allows, where appropriate, the emptying or filling of the storage tanks. This was based on the Korean philosophy that all people depend on the same water. It was also based on the concept of watershed management. River basin management parameters included low carbon emissions, growth of green areas, water self-suffiency and climate change adaption. Finally any development should ensure that the water system is the same after development as before.

Implementation of these changes also involved an amendment to design guide-lines. The Korean Ministry of Environment proposed a law that all the government buildings collect rainwater from the roof. The South Korean National Emergency Management Agency in their Natural Disaster Law, insisted rainfall runoff reducing facilities be installed. Selected sites were to serve as demonstration sites with full scientific and meteorological monitoring. There was also a Rain City: Regulations in Cities incentive which saw Seoul City apply the first proactive rainwater regula-tions to install rainwater tanks for new buildings and monitor the water level. The latter were accompanied by incentives and subsidy programs.

These initiatives led to the adoption of the Rain City policy by many cities throughout South Korea. Based on this many cities in Korea adapted this policy implementing and enforcing similar rainwater regulations. This was especially the case after recent drought problems. There is a movement toward the RainCity, where rainwater is stored for multipurpose use, instead of letting it run to surface water/sewerage drains.

Star City, Seoul :Star City was a major development project in Gwangjin-gu, a district in eastern Seoul. It consisted of more than 1300 apartment units. The design team for the complex, uniquely, comprised of an alliance of academics, the land owner and developer, the local government, the designer of the development proj-ect, and the general contractor. The Star City rainwater harvesting project was designed to capture the first 100 mm of rainfall on the complex and to use the har-vested water for toilets and gardening. Storage tanks were also installed of suffi-cient volume to mitigate flooding during the monsoon season. These tanks collected rainwater and groundwater. Rainwater used in the garden was infiltrated and returned to the Storage tank. Results from the Star City project show a local rain-water utilisation rate of 67%. This is a measure of the amount of rainwater used compared to the total annual rainfall in the area. Therefore 67% of the annual rain-fall was captured and utilised. The motto in Star City is "from Drain city to rain city by training Brain citizens".

This decentralised NBS water management system is replicated in the rain cities of Korea. It shows the importance of cooperation across the sectors, most particu-larly the involvement of local government (Han 2009).

6.5 Hybrid Infrastructure: Case Study at Regional Level

At the regional level, or the river basin level, a similar methodology applies. This involves a wide range of professional input, community involvement and natural engineering. However often times river basins cross national and international boundaries and present the added problem of a historical legacy in terms of disputed territory and bitter conflict. The application to river basins within a nation is a sim-pler task, if the implementation of water resources management can ever be classed as simple.

Philadelphia – Clean Waters, Clean City Program, – example of watershed management using an NBS Portfolio

The Philadelphia Water Department (PWD) provides water and wastewater treatment to a population of over 1.4 million people on the eastern seaboard of the USA. PWD has implemented a new approach to rainwater entitled "Green City, Clean Waters" (Philadelphia Water Department 2011). The Key to this approach is considering rainwater as an urban asset to be captured and used. The approach also considers the traditional concept of moving rainfall downstream as quickly as possible causes a reduction in the quality of the rainwater and increases the risk of flooding. The traditional concept is therefore not sustainable. The "Green City, Clean Waters" approach regards the participation of local communities as partners within the design, construction and management process as having equal importance with technological considerations. The approach stresses the importance of a knowledgeable community as a capable partner, able to actively participate in the design, construction and management of their own local infrastructure. Education at schools, community and business levels, has been fundamental to the water strategy. Initially the concepts were opposed by all levels of the engineering profession. However, now it is held as a case study in sustainable urban design.

The basic principles of the City's Green City, Clean Waters approach to regarding rainwater as a resource is the recycling of the harvested rainwater, its re-use, and infiltration of the rainwater into groundwater aquifers rather than piping it away from communities into stressed tributaries. This approach also maintains and upgrades water infrastructure. The strategy is aimed at reducing the storm water discharging to sewers, and using this water to improve the quality of the cities impervious areas by changing the landscape. PWD measure the progress of the approach through the number of Greened Acres achieved. Each Greened Acre refers to an acre of impervious cover within the combined sewer service area of Philadelphia that has at least the first inch of runoff managed by greened stormwater infrastructure. This includes the area of the stormwater management feature itself and the area that drains to it. If the land is impervious, it all runs off into the sewer and becomes polluted. A Greened Acre will stop 80–90% of this pollution from occurring.

Another feature of the Green City, Clean Waters concept is the integrated watershed Planning Approach. The city lies in the downstream section of a number of watersheds. Therefore without the involvement of upstream neighbours, stakeholders and local government agencies the effect of the approach would be limited. Watershed management involves the City and surrounding areas in a programme that aims to protect drinking water supplies, recreational sites and other water resources such as streams and parks.

The Green City, Clean Waters approach is based on a technology portfolio. In Philadelphia the NBS technology applied includes storm water tree trenches, downspout planter areas, green roofs, storm water planter areas, storm water wetlands, rain gardens. Other NBS innovations include green car parks, green schools, green homes and industries. In some cases these NBS technologies are used with pervious paving, rain water storage tanks and barrels and extended drainage facilities. When these non

NBS technologies are used in conjunction with NBS technologies it is referred to as an integrated or hybrid NBS system (http://www.phillywatersheds.org/).

6.6 Hybrid Infrastructure: Case Study at Global Level

At the global level the adoption of the methodology proposed in this publication involves political, theoretical and even philosophical issues. The input of the community to any NBS water resources strategy is a fine ideal but in reality the community is often the last bastion of resistance to the water resource management policies of local, regional and national government, as well as to business interests. The change of attitude, to viewing the local people, be they indigenous groups or internal migrants as custodians of their land is seismic. In effect it empowers local residents and gives them a large input into the decision making process. The cultural, political and demographic implications of such a change in focus are beyond the terms of this publication. Some areas of the world face situations where water is not going to be available in sufficient supplies to allow development of resources, and even in some cases, tolerable human living conditions. It is in this context that water has become a sustainability issue and the minimisation of water use has become a central tenet of sustainability.

However, this involves such themes as water recycling, massive infrastructure repairs, conservation and reclamation of destroyed water systems amongst others. It involves sustainable agriculture instead of industrial agriculture and local water rights for all. It involves strong laws and strong law enforcement against pollution and polluters, and transparency in all parts of the legal system. It involves equal access to the legal system also. Further considerations involve limits on industrial growth, the promotion of locally appropriate technology, an end to the construction of large dams and in order to offset ground water problems in the future, severe limits on groundwater extractions. Some of these topics are political, some are legislative, but most are beyond the scope of this publication.

Singapore – example of water underlying every government policy
Singapore is an example of the holistic approach to water, on the level of a nation state. All hard surfaces are considered water catchments, and all water that falls is considered a useful resource. These unprotected water catchments supply raw water which is treated to potable water standards. Wastewater is recycled and treated to drinking water standard and resold to users as a high quality brand "NEWater".

Singapore is an example where innovation in thinking has resulted in it moving from being a net importer of water to becoming self-sufficient in a period of 30 years. At the time of independence, Singapore was dependent upon importing water from its neighbour, Malaysia, to supplement water supply. This led to a policy decision in the Prime Minister's office "that water should govern every Government decision" (Num 2017).

Singaporeans refer to the water loop that is the basis for a sustainable water supply as the "Four Taps".

Singapore four taps water management strategy

Rainwater Harvesting	**Imported Water**	**NEWater**	**Desalination**
- collected from unprotected and protected catchments	- Malavasia	- Reclaimed water for non potable industry	- Reclaimed - Seawater (future)

   The First National Tap is potable water that comes from a network of waterways and reservoirs throughout the peninsula. These were initially fed by protected catchments, areas of land where industrial and housing developments were strictly controlled to protect the quality of the rainwater which was harvested. However as land demands increased for housing and industry, less land could be spared for additional protected water catchments. Therefore, the radical decision was taken to develop unprotected catchments. Unprotected catchments are areas of water catchment where all types of land use are allowed upstream of the storage area, regardless of potential effects on water quality. These include parks, pavements, roads drains etc. Every drop of rain is captured and all surfaces are regarded as water catchments. Water is designed into a system, as against being designed out via storm water drainage.

Example of rainwater collected from unprotected water catchment

   The Second National Tap refers to the imported water from Malaysia, which was the main source of water prior to the 1960s and is still purchased, but no longer has the same strategic value. The Third National Tap is NEWater. In 1998 two Singaporean engineers were sent on a study trip to the USA, specifically Southern California and Florida. This trip was a turning point in Singapore's efforts to recycle its wastewater. Wastewater is now treated to a potable standard and the end product is branded as "NEWater" and sold to Information Technology companies and also used in the national water supply. The Fourth National Tap is Desalination, which is being developed with the focus on reducing the power inputs required.

Wastewater treated to potable standards and rebranded "NEWater"

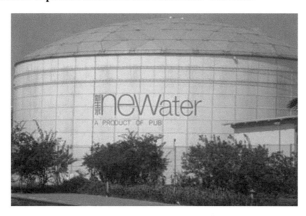

These initiatives were not without their difficulties. The engineering profession opposed the introduction of unprotected catchments on the grounds that water quality would be compromised and that drinking water standards would prove impossible to achieve. To overcome consumer reluctance to accept the concept of drinking water produced from sewage—literally their own excreted waste—a public campaign was conducted to convince consumers of the potability, lack of taste and odour of this treated water. This was rebranded "NEWater" to offset the "yuck factor". The campaign culminated with a high-visibility event at the 2002 National Day Parade when the then Prime Minister Tong lead sixty thousand people in a toast to Singapore with "NEWater" as the beverage.

More recently the Singapore government launched a campaign that they call ABC, referring to an active beautiful and clean waters programme. This is an initiative aimed at improving the quality of water and life by harnessing the full potential of waterbodies. By integrating the drains, canals and reservoirs with the surrounding environment in a holistic way, the ABC Waters Programme aims to create beautiful and clean streams, rivers, and lakes with postcard-pretty community spaces for all to enjoy. The idea is to promote local people's participation in, and use of, the waterways of Singapore in the hope that this will instill appreciation and water values into the community. Singapore is promoting the Worth of Water. As they say themselves, "we used to keep the community away from our water. Now we want them to use it, play in it, respect it. In other words, to take ownership of water and water resources".

References

Alfranca O, Garcia J, Varela H (2011) Economic valuation of a created wetland fed with treated wastewater located in a peri-urban park in Catalonia, Spain. Water Sci Technol 63(5):891–898

Balian E (2014) Outputs of the strategic foresight workshop "Nature based solutions in a BiodivERsA context", BiodivERsA report, Brussels, 11 June 2014. Available via http://www.biodiversa.org/687/download. Accessed 12 Sept 2015

Borjesson P (2008) Assessment of energy performance in the life-cycle of biogas production. Biomass Bioenergy 30(3):254–266

Brink P, Mazza L, Badura T, Kettunen M, Withana S (2012) Nature and its role in the transition to a green economy. Available via www.teebweb.org/wp-content/2012/10/Green-Economy-Report.pdf. Accessed 2 Dec 2015

Brix H (2004) Danish guidelines for small scale concructed wetland systems for onsite treatment of domestic sewage. In: Proceedings of the 9th international conference on wetland systems for water pollution control, Avignon, vol 9, pp 1–8

Brix H (2006) Onsite treatment of wastewater in willow systems. PhD Course, Use of Wetlands in water pollution control. International school of aquatic sciences, Arhus

Bruntland Commission (1987) Our common future. Oxford University Press. Available via http://www.un-documents.net/our-common-future.pdf. Accessed 10 Oct 2017

BS 8525-1 (2010) Greywater systems – codes of practice. British Standards Institute

BS8515 (2009) Rainwater harvesting systems – codes of practice. British Standards Institute

California Department of Public Health (2009) Water recycling criteria. California Code of Regulations. Available via http://www.cdph.ca.gov/certlic/drinkingwater/Documents/Lawbook/RWregulations-01-2009.pdf . Accessed 4 Nov 2015

Carson R (1962) Silent spring. Houghton Mifflin/Riverside Press, Boston/Cambridge, MA, QH545 P4C38

Cooper P, Job D, Green B, Shutes R (1996) Reed beds and constructed wetlands for wastewater treatment. WRc Publications, Medmenham, Marlow, May 2016

Council Directive 2000/60/EC of 23 October 2000 establishing a framework for Community action in the field of water policy. Official Journal of the European Communities L327 22.12.2000. Available via http://data.europa.eu/eli/dir/2000/60/OJ. Accessed 10 Dec 2015

De Vriend HJ, Van Koningsveld M (2012) Building with Nature: thinking, acting and interacting differently. EcoShape, Building with Nature, Dordrecht. Available via http://www.ecoshape.nl. Accessed 10 June 2015

Deltares (2015a) Building with Nature in the city. http://www.buildingwithnatureinthecity.com. Accessed 10 Aug 2016

Deltares (2015b) Building with Nature. Available via https://publicwiki.deltares.nl/display/BWN1/Building+Block+-+Inland+shores. Accessed 25 July 2015

© Springer International Publishing AG, part of Springer Nature 2018
S. O'Hogain, L. McCarton, *A Technology Portfolio of Nature Based Solutions*,
https://doi.org/10.1007/978-3-319-73281-7

Deshmukh S, Steinbergs C (2006) Engineers report on groundwater conditions, water supply and basin utilization in the orange county water district, Orange County Water USA

Directive 2014/52/EU of the European Parliament and of the Council of 16 April 2014 amending Directive 2011/92/EU on the assessment of the effects of certain public and private projects on the environment text with EEA relevance. Available via http://data.europa.eu/eli/dir/2014/52/OJ

EC (2011) Communication from the Commission to the European Parliament, the Council, the European Economic and Social Committee and the Committee of the Regions. Roadmap to a resource efficient Europe. Available via http://eur-lex.europa.eu/legal-content/GA/ALL/?uri=COM:2011:0571:FIN. Accessed 10 Dec 2015

ENSAT (2012) Enhancement of soil aquifer treatment. Technical final report, December 2012. Available via http://ec.europa.eu/environment/life/project/Projects/index.cfm?fuseaction=home.showFile&rep=file&fil=LIFE08_ENV_E_000117_FTR.pdf. Accessed 2 Sept 2015

Environmental Protection Agency (EPA) (2016) Report No. 161, Assessment of disposal options for treated wastewater from single houses in low permeability subsoils. Environmental Protection Agency, Ireland. Available via http://www.epa.ie/pubs/reports/research/water/researchreport161.html. Accessed 2 June 2016

European Communities (Good Agricultural Practice for Protection of Waters) Regulations (2010) (S.I. No. 610 of 2010). Available via https://www.agriculture.gov.ie/media/migration/ruralenvironment/environment/nitrates/SINo610of2010140111. Accessed 8 Dec 2015

European Communities (2014) Horizon 2020 Societal Challenge 5. Climate action, environment, resource efficiency and raw materials. Advisory group report. Available via http://ec.europa.eu/transparency/regexpert/index.cfm?do=groupDetail.groupDetail&groupID=2924. Accessed 8 Dec 2015

European Communities (2015) Towards an EU research and innovation policy agenda for nature-based solutions & re-naturing cities. Final report of the Horizon 2020 expert group on nature-based solutions and re-naturing cities. Available via http://bookshop.europa.eu/en/towards-an-eu-research-and-innovation-policy-agenda-for-nature-based-solutions-re-naturing-cities-pbKI0215162/. Accessed 8 June 2015

Grant G (2012) Ecosystem services come to town: greening cities by working with nature. Wiley Blackwell, London

Guo Z, Xiao X, Li D (2000) An assessment of ecosystem services: water flow regulation and hydroelectric power production. Ecol Appl 10(3):925–936

Gurluk S, Rehber E (2008) A travel cost study to estimate recreational value for a bird refuge at Lake Manyas, Turkey. J Environ Manag 88:1350–1360

Han M (2009) Promotion of rain cities in Korea-policies and case studies. Paper presented at the The International Rainwater Catchment Systems Association conference, Kuala Lumpur, 3–6 Aug 2009

Han M (2011) Urban planning for smart water management. Paper presented at ASPIRE smart water workshop, Tokyo, 4 Oct 2011

Kayser K, Kunst S (2002) Decentralised wastewater treatment in rural areas. Springer, Berlin/Heidleberg, pp 137–182

Limnos (2016) Limnos Ltd, personal communication 2016

Llorens E, Matamoros V, Domingo V, Bayona M, Garcia J (2009) Water quality improvement in a full-scale tertiary constructed wetland: Effects on conventional and specific organic contaminants. Sci Total Environ 407:2517–2524

Matamoros V, Garcia J, Bayona J (2008) Organic micropollutant removal in a full-scale surface flow constructed wetland fed with secondary effluent. Water Res 42:653–660

McCarton L, O'Hogain S (2011) Rainwater harvesting monitoring report. Carowholly National School, Westport. Mayo, Department of the Environment, Heritage and Local Government. Available via www.arrow.dit.ie. Accessed 2 Dec 2015

Mojca KK, Jerneja K, Ana VK, Nina G, (2011) Ecoremediation Educational Polygons in Slovenia as Good Examples of Experiential Learning of Geography. Literacy Information and Computer Education Journal 2 (3):481–490

Num P (2017) Smart water-Singapore case study. Puah Aik Num, Deputy Director, Technology and Water Quality Office, Pub, Singapore. Available via https://www.pub.gov.sg/. Accessed Jan 2017

O'Hogain S, McCarton L (2011) A review of zero discharge wastewater treatment systems using reed willow bed combinations in Ireland. In: Proceedings of 12th IWA international conference on wetland systems for water pollution control, Venice, 4–8 Oct 2010. Available via http://arrow.dit.ie/cgi/viewcontent.cgi?article=1029&context=engschcivcon. Accessed 5 June 2015

Perttu (1999) Environmental and hygienic aspects of willow copice in Sweden. Biomass Bioenergy Volume 16:291–297

Philadelphia Water Department (2011) Amended green city clean waters. Available via http://www.phillywatersheds.org/doc/GCCW_AmendedJune2011. Accessed 11 Nov 2014

Potschin M, Kretsch C, Haines-Young R, Furman E, Berry P, Baró F (2015) Nature-based solutions. OpenNESS Ecosystem Service Reference Book. Available via http://www.openness-project.eu/library/reference-book. Accessed 8 Dec 2015

SWRCB (2009) California state water resources control board recycled water policy. Retrieved July 2012. Available via http://www.swrcb.ca.gov/water_issues/programs/water_recycling_policy/. Accessed 10 Dec 2015

Uggetti E, Ferrer I, Arias C, Brix H, Garcia J (2012a) Carbon footprint of sludge treatment reed beds. Ecol Eng 44:298–302

Uggetti E, Garcia J, Lind S, Pertti J, Martikainen J, Ferrer I (2012b) Quantification of greenhouse gas emissions from sludge treatment wetlands. Water Res 46:1755–1762

US EPA (2012) Guidelines for water reuse. Available via http://nepis.epa.gov/Adobe/PDF/P100FS7K.pdf. Accessed 4 Nov 2015

Vymazal J (2005) Horizontal sub-surface flow and hybrid constructed wetland systems for wastewater treatment. Ecol Eng 25:478–490

Weedon C (2003) Compact vertical flow constructed wetland systems. Water Sci Technol 48(5):15–13

World Health Organization (WHO) (2006) WHO guidelines for the safe use of wastewater, excreta and greywater. United Nations Environment Program, Paris. Available via http://www.who.int/water_sanitation_health/wastewater/gsuww/en/. Accessed 10 Sept 2015

WssTP (2009) Water reuse, research and technology development needs for irrigated agriculture. Internal publication WssTP. Available via www.wsstp.eu. Accessed 8 Dec 2015

Index

© Springer International Publishing AG, part of Springer Nature 2018   119
S. O'Hogain, L. McCarton, *A Technology Portfolio of Nature Based Solutions*,
https://doi.org/10.1007/978-3-319-73281-7

Printed in the United States
By Bookmasters